KT-472-600

BASIC MOLECULAR AND CELL BIOLOGY

Second edition

BASIC MOLECULAR AND CELL BIOLOGY

Second edition

Published by the BMJ Publishing Group

Tavistock Square, London WC1H 9JR

First published 1988
Reprinted 1988
Second edition 1993

British Library Cataloguing in Publication Data.
A catalogue record for this book is available from the British Library.

ISBN 0-7279-0772-7

Typeset, printed and bound in Great Britain by
Latimer Trend & Company Ltd, Plymouth

Contents

Molecular and cell biology in clinical medicine: Introduction

David Weatherall

In the five years since this book was first published there has been remarkable progress in the application of molecular and cell biology to medical research. This new edition outlines some of these developments, with particular emphasis on their application to the study of human disease. Although, as in the previous edition, for convenience of presentation it is divided into two parts, one covering molecular biology and the other cell biology, the two fields are inseparable; molecular biology describes the anatomy and organisation of the molecules of living organisms, and modern cell biology encompasses the way in which they work together as an orchestrated whole to mediate and regulate cellular function.

By way of an introduction to this important and timely collection I shall summarise briefly the significance of some of the recent technical developments in this discipline together with the clinical relevance of some of the ground that is covered.

Technical advances

Since the first edition of this book there have been several important technical advances in molecular biology, which have already made important contributions to human molecular genetics. Perhaps the most important, and which for this reason has merited a new chapter, is the polymerase chain reaction (PCR). This valuable technique makes it possible to amplify small amounts of DNA in a very short time and has formed the basis for a whole variety of new approaches, which are particularly valuable for medical research. In particular it has been possible to use this method in the clinic for the rapid identification of genetic disease in fetal DNA and it has been modified to provide new and simpler approaches for DNA sequencing.

Rapid progress has also been made in defining highly polymorphic regions of DNA, including so called minisatellite and microsatellite DNA. This has been of particular value for identifying linkage markers, notably in mice, rats, and humans. Indeed, the generation of a linkage map of the whole of the human genome is well advanced and it is predicted, perhaps rather optimistically, that we may have a physical map of the entire human genome by the end of the second decade of the next century. With this in mind a great deal of effort is being put into developing automated methods for sequencing, which include the use of robotics.

These new developments, which are grouped under the heading of the human genome project, should greatly facilitate the medical applications of recombinant DNA technology. Not only will they make it easier to find genes for monogenic disorders but they should greatly facilitate the identification of some of the important genes that are implicated in the polygenic systems that underlie heart disease, hypertension, diabetes, and many more of our major killers.

Some applications of molecular biology in clinical medicine

The most immediate medical application of recombinant DNA technology has been in clinical genetics, in which it has been possible to work out the molecular pathology of many single gene diseases and to institute new techniques for identifying carriers and for prenatal diagnosis. We now have a very good idea of the repertoire of mutations that underlie single gene disorders, although there are still some surprises. For example, the recent discovery of the cause of the fragile X syndrome and several other related genetic disorders, which result from the amplification of short lengths of DNA either close to or within particular genes, underlines the extraordinary diversity of the molecular pathology of genetic disease.

In the first edition of this book we predicted that the application of reverse genetics—that is, finding genes for disorders of unknown aetiology by genetic linkage, sequencing them, and then predicting the structure of the protein product of the mutant gene—would be an invaluable tool for human genetics. This has turned out to be the case. This approach, which has now changed its name to positional cloning, has been responsible for the discovery of several medically important genes, notable those that are associated with cystic fibrosis and muscular dystrophy. The remarkable power of this discipline is evidenced by the extraordinarily rapid progress in understanding the pathology of cystic fibrosis after the relevant gene was discovered. Within two years, over 100 different mutations had been discovered, the function of the gene product as a chloride channel had been defined, at least in outline, and thoughts are already turning to gene replacement treatment.

Gene therapy, mentioned only briefly in the first edition of this book, has progressed considerably over the past five years. Several retrovirus vectors have been constructed, and methods have been worked out for ensuring the safety of this form of gene transfer. Other viral vectors have been developed and there has been some progress towards using nature's way of swapping genes, that is by site-directed recombination. But many problems remain. For example, in the case of bone marrow it is still difficult to transfect a sufficient number of haemopoietic stem cells to make the cure of a genetic disease likely. There are still concerns about the safety of using retroviral vectors, and there are problems to be overcome in obtaining adequate levels of expression of genes that have been inserted into foreign cells. But there have been some encouraging advances, and there is no doubt that gene therapy will be established, at least for a few disorders, over the next few years.

In introducing the last edition of this book I predicted that it would be in the study of the common illnesses of Western society, such as degenerative vascular disease, diabetes, psychiatric disorders, cancer, and rheumatism, that molecular genetics might in the long term have its most important role. In particular, by defining some of the important genes in the polygenic systems that underlie our common diseases we might possibly learn more about their underlying cause. There has been some genuine progress in research in this topic although, not surprisingly perhaps, there have been false leads as well as successes. Perhaps the most spectacular progress has been made towards an understanding of the genetics of both type I (insulin dependent) and type II (non-insulin dependent) diabetes. Some of the major players in the genetics of vascular disease have started to emerge, perhaps the most surprising and novel being the recent discovery that a polymorphism of the gene that regulates angiotensin converting enzyme (ACE) shows strong association with coronary artery disease.

But the most spectacular progress towards an understanding of the pathogenesis of our common killers over the past few years has come from work in the cancer field. There have been major advances in our understanding of the role of mutations of oncogenes in the generation of cancer, and the role of anti-oncogenes has been clearly defined. Work on colon cancer has given some indication of the number of different mutations that may be required to generate a common malignancy, and recent studies on the genetics of breast cancer suggest that one particular gene, on chromosome 17, may be a major player.

Cloned genes are already finding wide application in the diagnosis of bacterial, viral, and parasitic illnesses, and probes for oncogenes and for immunoglobulin and T cell receptor genes are being used to type tumours. Several human genes have been persuaded to synthesise their products in bacterial systems. The correction of the anaemia of renal failure by genetically engineered erythropoietin is a good example of an early success in the rapidly expanding biotechnology industry. Numerous other therapeutic and diagnostic agents are under evaluation. The problem of resistance to antibiotics is being approached by producing hybrid antibiotics, constructed by altering the genes responsible for antibiotic production in fungi and bacteria. Recombinant DNA technology offers new approaches to vaccine production—for example, augmenting vaccinia virus with genes from other viruses such as rabies, and the production of large quantities of specific antigens from malarial parasites. The discovery of the retrovirus that causes AIDS so soon after the condition was first recognised exemplifies the strength of this new technology; if a means to control AIDS is found it is likely to come from the same discipline.

Amalgamation of molecular and cellular biology

Although less newsworthy, advances in our understanding of cell biology have been almost as rapid as those in molecular biology. This subject has equal potential for medical application, particularly for understanding cancer, autoimmune disease, human development, and the genesis of congenital malformation.

The application of recombinant DNA technology to the problems of cell biology is starting to unravel some of the mysteries of cell behaviour. For example, in many tissues there are pluripotential stem cells that go through cycles of multiplication before entering terminal differentiation towards their specialised functions. It is becoming possible to define specific genes implicated in the regulation of different phases of the cell cycle and in the critical steps of commitment and differentiation. The availability of gene probes and methods of studying the physical properties of inactive and active genes is enabling many of the processes of proliferation and differentiation to be described at the molecular level.

In understanding the orchestration of cellular proliferation and function it is important to learn how cells respond to external regulatory signals, such as hormones and growth factors. Considerable progress has been made in relating the action of external regulatory molecules to the cyclic AMP, inositol triphosphate (ITP), and protein kinase C (PKC) messenger systems dependent on phospholipids. Equally important advances have been made in purifying some of the growth factors and other regulatory proteins that control cell proliferation. Once their structure is known their genes can be isolated and, by recombinant DNA technology, used to produce large quantitites of their products for investigating how they work and for assessing their clinical potential. A whole family of haemopoietic growth factors has been obtained in this way, many of which are already available for analysing how they regulate marrow progenitors and for use in clinical trials.

An understanding of what controls cell proliferation and differentiation has important applications for clinical practice. As mentioned earlier, the discovery of oncogenes, genes that are normally concerned with various cellular housekeeping activities, and their abnormal function in cancer has opened up a whole new subject of investigation into the basic processes of malignant transformation. The finding that abnormal activation of these genes is often related to the chromosomal changes that are found in many common cancers has brought together cell biology and cytogenetics in an exciting new facet of cancer research. The message from these studies is that cancer is a diverse disorder resulting from genetic damage of many kinds, including recessive or dominant mutations, major DNA rearrangements, and point mutations, all of which may disturb the expression of biochemical function of the affected genes. We now seem to have the tools with which to unravel these processes.

Studies of cell membrane and receptor function promise to have equally important clinical applications. For example, work on the lifestyle of the low density lipoprotein (LDL) receptor, including its internalisation and recycling and how these are regulated, has helped to define the many stages at which cholesterol metabolism may be altered, and has offered us new possibilities for the pharmacological control of cholesterol concentrations. By monoclonal antibody and protein engineering technology it will probably be possible to modify receptor function. The isolation of the T cell receptor and studies of the cell biology of antigen recognition have opened up a completely new approach to sorting out the mechanisms of the autoimmune processes that may underlie many common diseases. The discovery and characterisation of the "supergene" family that regulates the immune system, and of the lymphokines that constitute its effector molecules, are other remarkable examples of the combined strengths of cell and molecular biology and their potential for the medical services.

The future

We hope that this collection of short articles will give clinicians some insight into the way in which the medical sciences may be moving over the next few years. In effect, we appear to be changing from the era of whole-patient physiology and pathology to the study of disease at the molecular and cellular level. This will undoubtedly alter the pattern of medical research. One effect will be to cloud the current compartmentalisation of medicine into watertight specialties; in the future cardiologists, nephrologists, and so on will be using the same techniques of cell and molecular biology to study diseases particular to their subject. It therefore follows that we may have to revise the way that research departments and resources are organised if we are to take full advantage of what recombinant DNA technology has to offer.

Clinicians may be worried about this reductionist approach to the study of disease, particularly at a time when we are being accused of ignoring the broader pastoral aspects of patient care and when there is a cry for a return to holistic medicine—that is, to rediscover how to be a good doctor. But there is no reason why a proper understanding of disease processes at the molecular and cellular level should have a deleterious effect on clinical care or on doctors' attitudes to their patients. On the contrary, in the long term it should enable us to treat many diseases more rationally and hence to reduce the proliferation of high technology medicine that characterises much of the current medical scene. But this will not happen overnight; there is likely to be a long period of development and evaluation of our new technology before its clinical application can be fully assesed. There may well be disappointments, but in the long term these new developments are likely to be of enormous benefit.

All this will require some rethinking of medical education and the way in which our clinical schools and postgraduate programmes are organised in the future; the profession will have to take on board and accommodate individuals with a much broader range of scientific skills than has been necessary hitherto. This will require some urgent thought by those who are concerned with training our doctors in the future. I hope that this book will help to provide some background information on which we can develop an informed debate on how these changes might be made while, at the same time, maintaining high standards of patient care.

Methods in molecular medicine

R K Craig

During the past decade an array of powerful new diagnostic techniques has been developed based on nucleic acid hybridisation and gene probes. These permit the direct analysis of genes in DNA extracted from the nuclei of human cells or, alternatively, of gene transcripts in the form of messenger RNA (mRNA), the template for protein synthesis found in the cytoplasm. These techniques contrast with the use of antibodies, which permit the analysis of the gene product or protein, either in the cell or in cell secretions—for example, plasma (fig 1).

The concept of hybridisation

DNA is made up of four building blocks or bases: adenine (A), guanine (G), cystosine (C), and thymine (T). Within a strand of DNA the bases are linked by a sugar-phosphate backbone. Within the cell the DNA is in a highly ordered double stranded helical structure. The helical structure is maintained through specific hydrogen bonding interactions between complementary bases, so that A in one strand always pairs, or hybridises, with T in the other and C hybridises with G. Thus A and T and C and G are termed complementary bases and must always be present in equivalent amounts in double stranded DNA.

A region of DNA that encodes a protein is termed a gene. The genetic information is encoded by the sequence of bases through a non-overlapping code in which three bases (a triplet) determine a particular amino acid (see reference 1 for well illustrated reading). For a gene to be expressed an enzyme, RNA polymerase II, copies or transcribes one strand of the DNA into mRNA, which is then decoded or translated by the protein synthesis machinery in the

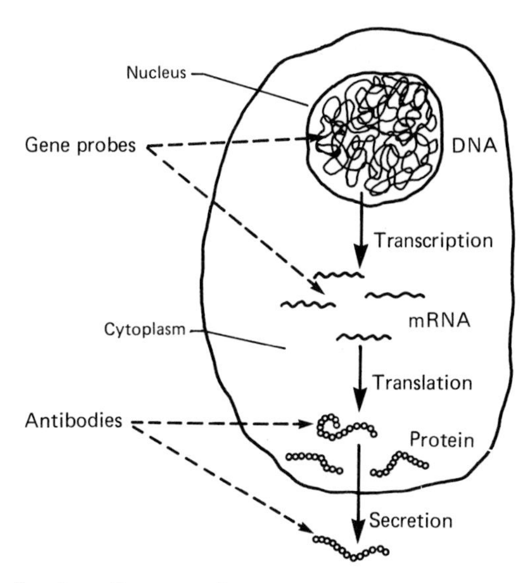

FIG 1—Site of action of gene probes.

cytoplasm. The mRNA comprises a single stranded polynucleotide chain with a sugar-phosphate backbone in which the order of bases is the complement of the transcribed DNA strand of the gene. In RNA, T is replaced by a closely related base, uracil (U), which also will form a base pair with or hybridise to A.

When DNA is isolated from cells or tissues it is usually recovered in a double stranded or native form. If a solution of DNA is heated or placed in an alkaline environment, however, the hydrogen bonds between the bases are broken and the strands separate, giving rise to denatured or single stranded DNA. If a heat treated denatured DNA solution is permitted to cool very slowly then the complementary strands will again form a base pair or hybridise, resulting in the formation of double helical DNA, indistinguishable from the original native preparation (fig 2). It is the ability of long stretches of one strand of denatured DNA, containing infinitely variable combinations of the four bases, to hybridise in a precise manner to its complementary strand that forms the basis of nucleic acid hybridisation technology. Thus, as we shall see, radiolabelled single stranded RNA or DNA sequences, complementary to the gene sequence of interest, can be used with great precision to identify the presence of individual genes in DNA preparations and specific mRNA in RNA preparations, or even within individual cells.

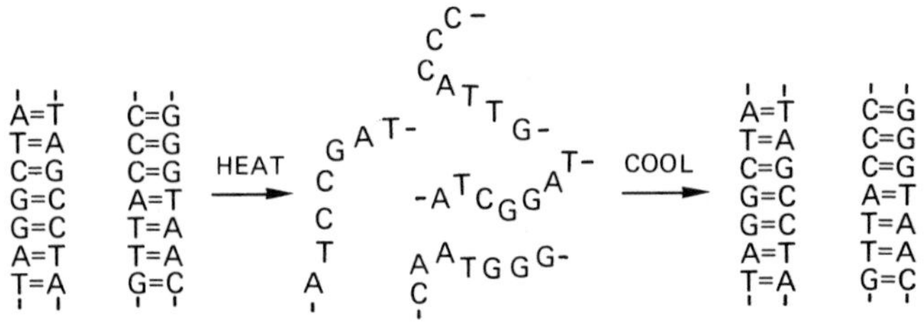

FIG 2—Effect of heating in separating double stranded DNA and of slow cooling in hybridising the strands to form double stranded DNA indistinguishable from the original.

Looking at genes

The potential coding capacity of DNA in a single human cell is enormous. In each diploid cell double stranded DNA is distributed between 23 pairs of chromosomes, comprising about $3{\cdot}5 \times 10^9$ base pairs, or two metres of DNA if stretched end to end. Potentially this might encode 2 million genes. In reality half of the DNA is composed of short simple sequences repeated many times—for example, (ACAAACT)n, termed satellite DNA. These sequences have no obvious function. Of the remainder, some comprise more complex sequences, which occur many times but at different points in the genome. The rest, the so called unique sequences, or single copy genes, represent long tracts of sequence many kilobases (1000 bases) long, each of which occurs perhaps only once in each haploid cell. These encode somewhere between 50 000 and 200 000 different protein products.

The analysis of a single gene within the vast amount of genetic material in a cell depends on the following principles. Firstly, the long and unmanageable human DNA isolated from cells and tissues must be cleaved into smaller but reproducible pieces. Secondly, hybridisation probes specific for individual genes must be generated. Thirdly, techniques are required which permit the direct visual comparison of, for instance, single copy genes from different individuals or genes from normal or cancerous tissue from an individual.

The identification of a class of endonucleases (restriction endonucleases), which cleave DNA at specific sites determined by a short sequence of bases, provided molecular scissors. With these a solution of native DNA, representative of DNA isolated from many millions of cells, can be cut so that the DNA from each cell is cleaved into an identical set of fragments. Different restriction

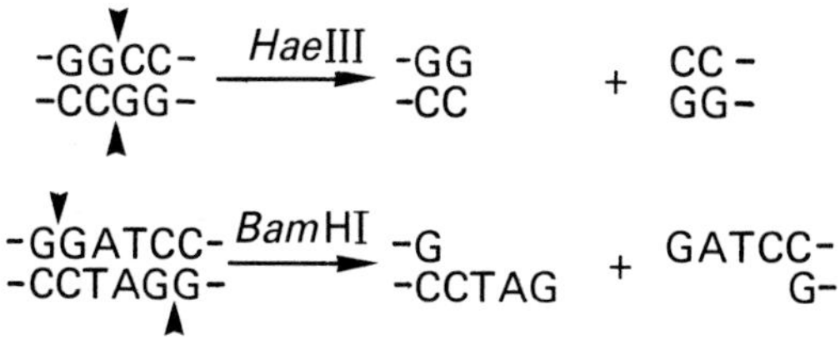

```
-GGCC-  HaeIII   -GG        CC-
-CCGG- -------->  -CC   +   GG-

-GGATCC- BamHI   -G         GATCC-
-CCTAGG- -------> -CCTAG  +     G-
```

FIG 3—Sequences cleaved by two restriction enzymes, *Hae*III and *Bam*HI.

enzymes digest, or cleave, double stranded DNA at different recognition sequences and therefore cleave a given DNA preparation into different fragments of different lengths, which are characteristic of the enzyme. Some enzymes recognise a sequence of four bases—for example, *Hae*III—others five or six bases—for example, *Bam*HI (fig 3). Enzymes recognising four bases will cleave the DNA more often than those recognising five or six bases. DNA fragments can then be separated electrophoretically on the basis of size, and the relative position of a specific gene can be determined by hybridisation (see below).

The molecular cloning and characterisation of human gene sequences is beyond the scope of this chapter (for further reading see reference 2), but large numbers of bacterial plasmids, each containing part of the human genome, may be constructed (see fig 4). Individual plasmids containing any desired gene may be selected from a "gene library", and the cloned or inserted gene sequence of interest may be characterised by DNA sequence analysis. Each characterised plasmid containing a selected gene sequence may then be grown (amplified) in bacterial host cells, and large amounts of plasmid DNA containing the inserted gene can be isolated. This recombinant plasmid DNA may then be used as a source of a specific gene probe in subsequent hybridisation analyses. Alternatively, once the nucleotide sequence of a gene is known, a desired sequence of bases may be synthesised chemically—an oligodeoxynucleotide—and used as a hybridisation probe.

The analysis of individual gene structure (gene mapping) in health and disease uses purified DNA isolated from tissues or, more often, peripheral blood lymphocytes. The isolated DNA is treated with a specific restriction enzyme, and the restriction fragments generated are separated on the basis of size by electrophoresis on an agarose gel. The separated DNA fragments are denatured in the gel by alkali treatment and then transferred on to

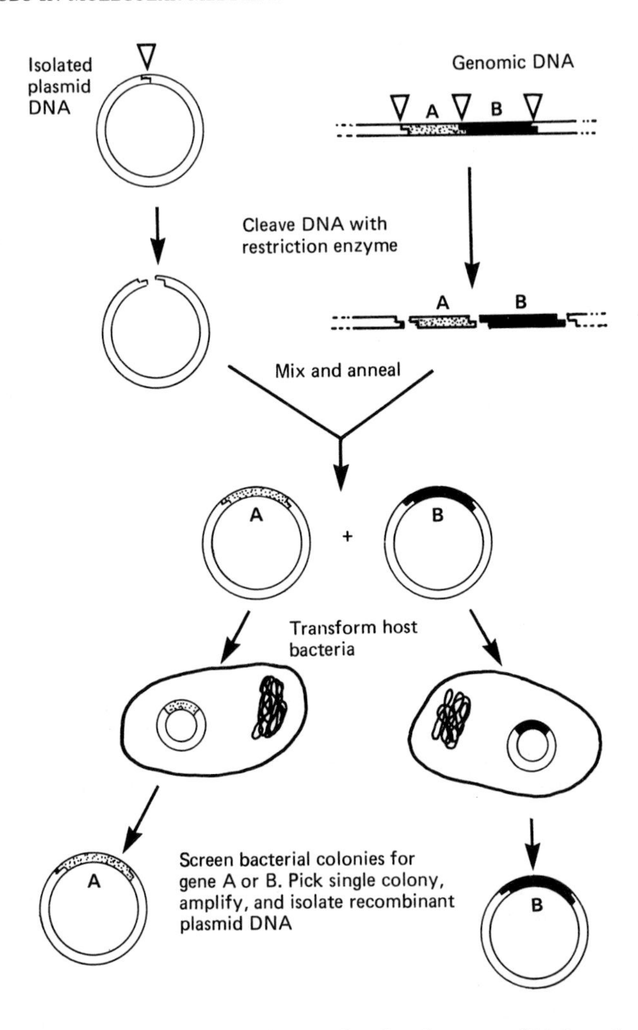

FIG 4—Generation of hybridisation probes by cleavage of high molecular weight DNA with restriction enzymes, followed by insertion of the individual fragments into bacterial plasmids.

a nitrocellulose membrane using a process called Southern blotting, named after the inventor, Professor E M Southern (fig 5A).

The size of a restriction fragment of DNA, which contains a specific single copy gene, may then be determined by hybridisation analysis using a radiolabelled gene probe complementary to the now denatured gene sequence under investigation. Radiolabelling permits the position of the hybridised gene probe to be pinpointed

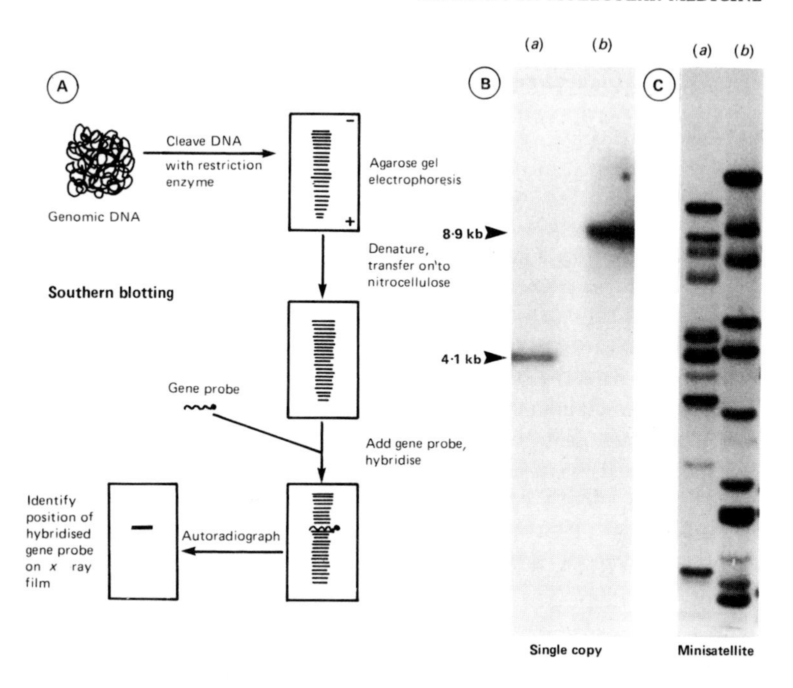

FIG 5—Looking at genes. (A) Southern blotting. (B) Identification of restriction fragments of human genomic DNA containing the calcitonin gene after digesting genomic DNA isolated from a single individual with the restriction enzyme (*a*) *Pst*I or (*b*) *Hin*dIII. Fragments were Southern blotted and those containing the calcitonin gene identified by hybridisation using a calcitonin specific gene probe. (C) Genetic fingerprints of two unrelated individuals (a) and (b) after digestion of their genomic DNA with the restriction enzyme *Hin*fI, Southern blotting, and hybridisation with a "minisatellite" gene probe.

by autoradiography—on *x* ray film—and consequently the size of the identified gene fragment can be determined relative to markers separated in parallel by electrophoresis. Thus the Southern blot in figure 5B shows two different sizes of restriction fragment containing the single copy human calcitonin gene after digestion of human DNA with two different restriction enzymes. Cleavage of the genomic DNA with different restriction endonucleases permits us to build up a picture (or map) of the distribution of restriction enzyme cleavage sites close to specific genes in the human genome. Southern blots may also be probed with sequences that occur many times and at different parts of the genome. These repeated or "minisatellite" sequences permit the visualisation of different parts of the genome on a single gel. Minisatellite probes provide a powerful means of detecting sequence differences in the genome of individuals, the differences generally reflecting the presence or otherwise of recognition sites for specific restriction enzymes (see fig 5C). Minisatellite probes have made a significant impact on forensic medicine (see reference 3).

Looking at gene expression

The technique of hybridisation analysis using radiolabelled cloned gene probes may also be used to identify specific RNA transcripts within cells and tissues. Total cellular RNA may be isolated from cells in culture or from fresh or frozen tissue, and the RNA species may be separated on the basis of size by gel electrophoresis and then transferred to a nitrocellulose membrane by blotting. RNA, or "northern," blotting follows the same principles as DNA blotting, except that there is no need for denaturing (RNA is single stranded) or cleaving with restriction enzymes. The presence and size of a specific mRNA species is then determined using a radiolabelled hybridisation probe followed by autoradiography. Northern or RNA blotting can be used to determine the presence or absence of mRNA species and the relative amount of these mRNAs in normal and diseased states—for example, the relative amounts of α_1 antitrypsin mRNA present in RNA from the liver of normal (MM) or α_1 antitrypsin deficient (ZZ) individuals (fig 6A).

Northern blotting will not, however, provide information on which cells within a tissue are expressing a gene of interest. This requires a technique analogous to immunocytochemistry, known as in situ hybridisation (see reference 4). This entails the hybridisation of radiolabelled gene probes complementary to the mRNA species in fixed tissue sections frozen or embedded in paraffin. After hybridisation the sections are dipped in photographic emulsion and stained. Cells that express the gene under investigation are then identified by the presence of silver grains localised over the cell. The example shown (fig 6B) identifies infiltrating B cells synthesising immunoglobulin light chain (κ) mRNA in a formalin fixed, paraffin embedded section of a breast tumour.

Since this chapter was first drafted, a single technique of remarkable simplicity, flexibility, and sensitivity has revolutionised our ability not only to look at genes, but now to examine routinely single gene defects at the level of a single base change or deletion. Moreover, such analysis can be performed rapidly and

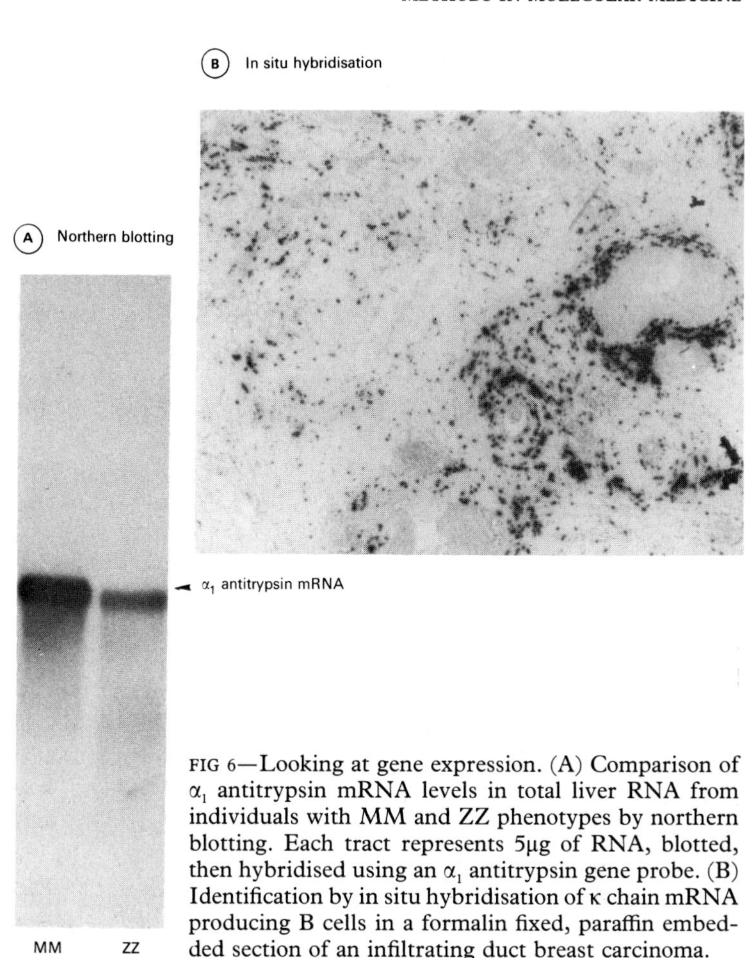

FIG 6—Looking at gene expression. (A) Comparison of α_1 antitrypsin mRNA levels in total liver RNA from individuals with MM and ZZ phenotypes by northern blotting. Each tract represents 5μg of RNA, blotted, then hybridised using an α_1 antitrypsin gene probe. (B) Identification by in situ hybridisation of κ chain mRNA producing B cells in a formalin fixed, paraffin embedded section of an infiltrating duct breast carcinoma.

routinely on genomic DNA isolated from just a few cells present for instance in saliva or a single hair follicle. The full impact of the polymerase chain reaction is reviewed by Professor A F Markham in a later chapter.

Conclusion

The power and sensitivity of the technology now available to those who wish to apply molecular techniques in clinical research are remarkable. Moreover, characterised gene probes are readily available, as are kits to radiolabel them. The technology is now commonplace and applicable in all branches of diagnostic medicine.

References

1 Darnell J, Lodish H, Baltimore D, eds. *Molecular cell biology*. Washington, DC: Scientific American Books, 1986.
2 Brown TA. *An introduction to gene cloning*. New York: Van Nostrand Reinhold, 1986.
3 Gill P, Jeffreys AJ, Werrett DJ. Forensic application of DNA fingerprints. *Nature* 1985;**318**:577–9.
4 Hoefler H, Childers H, Montminy MR, Lechan RM, Goodman RH, Wolfe HJ. In situ hybridization methods for the detection of somatostatin mRNA in tissue sections using antisense RNA probes. *Histochem J* 1986;**15**:597–604.

Impact of molecular biology on clinical genetics

Marcus Pembrey

The central activity of clinical genetics is genetic counselling, the aim of which is to provide information about the risks to offspring at a time appropriate to considering the options available for modifying the outcome and to put those risks in perspective. The objective of genetic services is not to reduce the birth incidence of genetic disorders, but to give the family an informed choice.

However, there is much evidence to show that, with serious or life threatening disorders, genetic counselling causes many couples to choose to modify their reproductive behaviour, even when the options available are far from satisfactory. Our ability to use DNA probes (labelled single stranded DNA fragments that bind only to specific genes or DNA sequences from the subject's total DNA) to analyse mutations or track the inheritance of genes through families has enormous potential as an aid to genetic counselling. Unlike protein analysis, DNA analysis is not dependent on examining tissue in which the gene in question is expressed. Couple DNA analysis with chorionic villus sampling in the first trimester of pregnancy and the effect for some families is nothing short of revolutionary.

DNA analysis was first used in the prenatal diagnosis of the haemoglobinopathies. Kan and Dozy reported the diagnosis of sickle cell anaemia on amniotic cells taken in the second trimester, and Old *et al* were first to report fetal diagnosis of β thalassaemia in the first trimester based on the analysis of chorionic villus DNA. Since then so many monogenic disorders have become amenable to this approach that it is not possible to give a comprehensive up to date list. The worldwide human genome project and related research in molecular medicine are committed to mapping and eventually sequencing all the expressed human genes in the next decade or so. The challenge for people providing clinical genetic services is to develop generally applicable methods that can quickly translate these basic discoveries into robust and reliable tests for prenatal diagnosis and carrier detection. The table lists the more common mendelian disorders. All those associated with a known single gene locus are amenable to genetic prediction by DNA analysis. For many, the demand is for early prenatal diagnosis. There is also a demand for carrier testing in females with the X linked disorders and presymptomatic testing for some of the late onset autosomal dominant disorders. No matter how rare the genetic disease, specific inquiries on behalf of a family at risk should always be made of the regional genetic service. The interval between gene localisation or cloning and the clinical application of this information can be very short, if adequate counselling and laboratory services are available. People caring for families at risk should be aware of what can be offered and of the importance of referral before a pregnancy. Families increasingly expect the options of carrier testing and prenatal diagnosis, and increasingly there are tests that can help them.

There are two fundamentally different approaches to genetic prediction by DNA analysis: detecting the mutation and gene tracking.

Detecting the mutation

The principle of detecting the mutation is the same as that of the traditional diagnostic test—namely, to detect the relevant differences from normal in the DNA sequence of the gene in question. The first requirement is a DNA probe capable of selectively binding (hybridising) to that part of the gene in which the mutation occurs. This immediately highlights the main limitation of this approach. In most of the conditions listed in the table, the site (and type) of mutation within the gene varies between families, and searching the whole gene for anything other than large deletions is currently very time consuming and only possible in a service context for just a few small genes, such as β globin, which is about 2000 base pairs in length. Haemophilia A can result from mutations more or less anywhere along the factor VIII gene, which consists of 186 000 base pairs. The dystrophin gene, implicated in Duchenne and Becker muscular dystrophy, is two million base pairs long. Thus, though the ideal is to detect the mutation

Common monogenic disorders

Disorder (chromosomal location of gene)	Rough birth incidence per 1000 in Europe
Autosomal dominant	
Familial hypercholesterolaemia (19p)	2·0
Huntington's chorea (4p)	0·5
Neurofibromatosis (17q)	0·3
Myotonic dystrophy (19q)	0·2
Hereditary motor and sensory neuropathy type I (17p)	0·2
Adult polycystic kidney disease (16p)	0·9
Familial adenomatous polyposis (5q)	0·1
Tuberose sclerosis (9q or 16p)	0·03
Retinoblastoma (13q)	0·05
Osteogenesis imperfecta (7q or 17q)	0·04
Marfan's syndrome (15q)	0·04
Dominant blindness	0·1
Dominant childhood deafness	0·1
Other dominants	2·0
Autosomal recessive	
Cystic fibrosis (7p)	0·5
Sickle cell anaemia (11p)	depends on ethnic origin
β thalassaemia (11p)	depends on ethnic origin
Phenylketonuria (12q)	0·1
Neurogenic muscle atrophies (5q)	0·1
Congenital adrenal hyperplasia (21 hydroxylase deficiency) (6p)	0·1
Recessive severe congenital deafness	0·2
Recessive blindness	0·1
Recessive non-specific severe mental retardation	0·5
Other recessives	2·0
X linked	
Duchenne muscular dystrophy (Xp21)	0·25 (males)
Haemophilia A (Xq28)	0·1 (males)
Haemophilia B (Xq27)	0·03 (males)
Ichthyosis (Xp22)	0·1 (males)
Fragile X mental retardation (Xq27)	0·75 (males)
Other X linked	0·6 (males)
	All serious mendelian disorders 10/1000 total livebirths

itself, this is still problematic in many families, even when the relevant gene is cloned. However, methods of DNA analysis are improving all the time. The most important technical development since the first edition of this book was written has been the application of the polymerase chain reaction (PCR) to DNA analysis. Once the nucleotide base sequence of a target region of

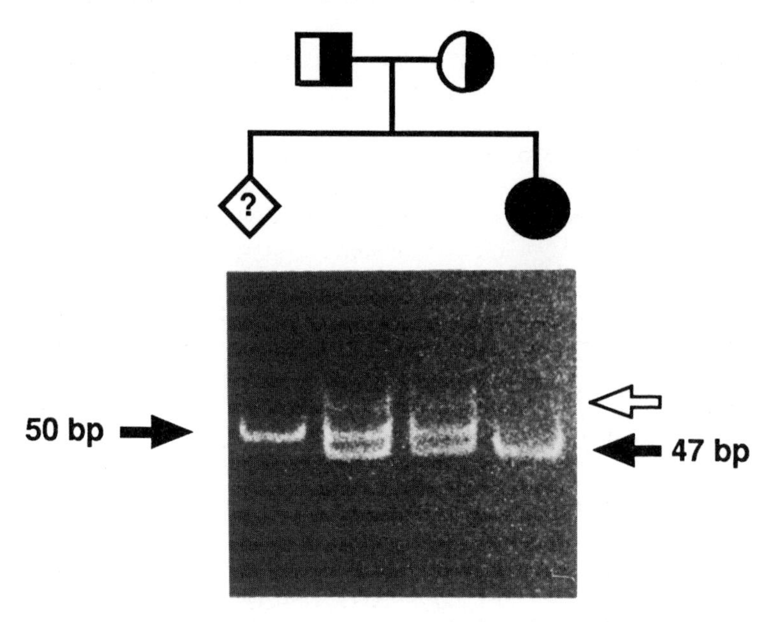

FIG 1—Pedigree showing a female with cystic fibrosis, her carrier parents, and a fetus undergoing prenatal diagnosis. Below each individual in the pedigree is the DNA track showing the normal 50 base pair DNA fragment and the 47 base pair fragment, due to deletion of codon 508. The open arrow shows a faint band in the heterozygotes due to a pairing between the 50 base pair and 47 base pair DNA strands (heteroduplex) which, because of its configuration, travels more slowly in the polyacrylamide gel. The fragments are generated by the polymerase chain reaction (PCR) using primers that flank codon 508 and are stained with ethidium bromide.

DNA is known, the polymerase chain reaction can be used to amplify selectively that length of DNA several million-fold in just a few hours. The principle is illustrated in the relevant chapter.

An additional benefit of the polymerase chain reaction is its exquisite sensitivity. With great care being taken to avoid contamination, DNA can be amplified from a single DNA molecule, and this has permitted the development of the single (or few) cell method of DNA analysis necessary for genetic tests on preimplantation embryos obtained during in vitro fertilisation.

The polymerase chain reaction allows the amplified fragments of DNA to be observed direct by staining the DNA only. Figure 1 shows the detection of the most common mutation in cystic fibrosis—a deletion of three nucleotide bases, codon number 508, which code for a phenylalanine in the normal gene product. In this analysis, electrophoresis on polyacrylamide gel can separate short amplified DNA fragments that differ by only three nucleotides in length.

The first point to make about mutation detection is that there are now routine methods to detect virtually any known mutant sequence, whether it be a point mutation, in which one nucleotide base is substituted for another, or some more substantial change in DNA sequence, such as large deletions, duplications, or some variable increase in the size of a repeating DNA sequence. The practical issue therefore becomes one of knowing which diseases are always or often caused by one or more particular small mutations as opposed to some gross deletion or rearrangement that, although perhaps variable between families, is nevertheless detected readily.

Sickle cell anaemia and α_1 antitrypsin deficiency (ZZ variant) are important examples of conditions in which you can expect the mutation to be the same in every affected family. Huntington's chorea may turn out to be another example because the fact that affected individuals generally reproduce before the clinical onset of the disease means that many families will have a common ancestor. Although there are about 60 different ways in which the β globin gene can mutate to produce β thalassaemia, one particular mutation often predominates within a given population, as has been shown clearly for subpopulations in the Mediterranean. A similar picture has emerged with cystic fibrosis. The codon 508 deletion accounts for about 88% of cystic fibrosis chromosomes in Denmark, 75% in Britain and northern Europe, 50% in Italy, but only about 30% in the eastern Mediterranean countries.

Readily detectable DNA deletions are the rule in Far East α thalassaemia and account for a little more than half of the mutations causing Duchenne muscular dystrophy and 30% of the severe congenital form of steroid 21-hydroxylase deficiency. A specific large duplication of DNA sequences on chromosome 17 seems to be the mutation in most patients with hereditary motor and sensory neuropathy type 1, and nearly all patients with fragile X mental retardation (FMR) have a variable increase in the size of a CGG repeat sequence at the beginning of the FMR 1 gene. Myotonic dystrophy is caused by a similar type of mutation—an expansion in a CTG repeat sequence at the end of the myotonin kinase gene. The repeat sequence expands with successive generations, which correlates with increasing severity of the disease. There now seems to be a molecular genetic explanation for a first generation with just cataracts but a third generation with congenital myotonic dystrophy.

When one specific mutation predominates, population screening for carriers of such X linked or autosomal recessive disorders becomes a realistic possibility. However simple, cheap, and reliable DNA analysis is, it is only one aspect of such screening programmes, and providing the associated counselling and educational services can be a major undertaking. Nevertheless, mutation detection can offer the prospect of alerting couples at risk before they have a severely handicapped child, as well as providing the early prenatal diagnosis that they may well choose as a means of achieving healthy children.

Gene tracking

Gene tracking asks the question: "Has this family member or fetus inherited the same relevant chromosome region(s) as a previously affected member?" There are variations to this question, but all depend on formulating a simple study with the family members available to establish the band on the autoradiograph of a gel with which the disease is coinheriting (fig 2). It is an approach that is independent of the particular type of mutation within the gene that is responsible for the disease, and herein lies its great usefulness. It requires either a gene specific probe or a chromosome region specific probe for a sequence known to be closely linked to the disease locus.

In gene tracking there has to be some way of distinguishing each chromosome of the homologous pair in key family members. This is done by exploiting naturally occurring variations of DNA sequence, known as restriction fragment length polymorphisms.

Restriction fragment length polymorphism

Only a small percentage of total genomic DNA is actually a coding sequence for proteins. The non-coding regions that flank genes, the intergenic DNA, and to some extent the intervening sequences, are conserved less during evolution, and point mutations or other DNA sequence variations are tolerated and become established in populations. Experience so far suggests that on average 1 in 150 nucleotide bases differ between the chromosome pair. A number of these DNA sequence polymorphisms concern the recognition sequence of four to eight bases of a particular restriction enzyme, and this results, on digestion, in different sized

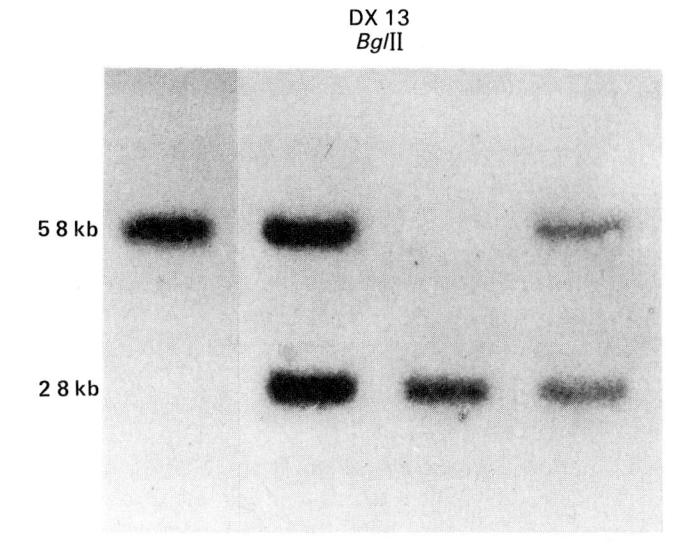

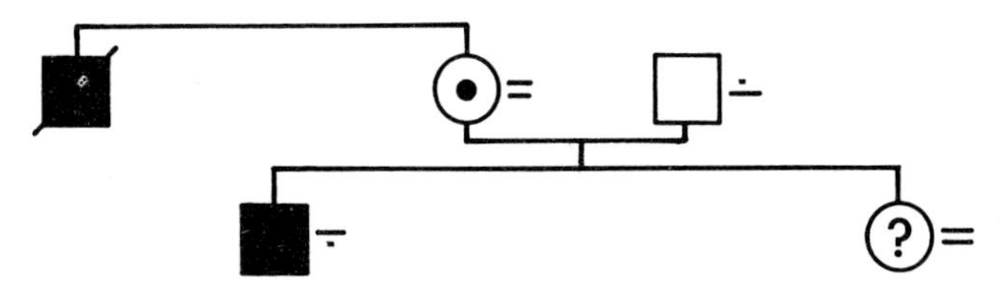

FIG 2—Autoradiograph of DNA tracks from four family members digested with restriction enzyme *Bgl*II and hybridised with probe DX13, which is closely linked to haemophilia A. The restriction fragment length polymorphism is represented by the polymorphic bands 5·8 kb and 2·8 kb. The 5·8 kb band appears to be coinheriting with the haemophilia mutation, and the sister has inherited this band from her mother and the 2·8 kb band from her father. Barring an error due to recombination, she is therefore a carrier.

restriction fragments from each chromosome of the homologous chromosome pair. Alternatively, a hypervariable region may exist between two restriction enzyme cutting sites, in which one chromosome has a different number of repeated sequences from the other and therefore has a restriction fragment of a different size. Thus a restriction fragment length polymorphism is a relatively common change in DNA sequence that either destroys or creates a restriction enzyme recognition site or alters the distance between two sites. In anyone who is heterozygous for a restriction fragment length polymorphism, one restriction band pattern corresponds to one chromosome and the other band pattern to the other chromosome of the pair. This allows us to track the transmission of a single chromosome region through a family and to see if a particular monogenic disease coinherits with the polymorphic site; in other words, we can perform classic linkage studies. Analysing dinucleotide, trinucleotide, and tetranucleotide repeat sequences by the polymerase chain reaction is proving valuable in gene tracking.

Figure 2 shows a restriction fragment length polymorphism that is detected by the probe DX13 after the DNA has been digested with the restriction enzyme *Bgl*II. DX13 detects a DNA sequence (by international agreement called locus DXS15), the function of which, if any, is not known but which is closely linked to the factor VIII gene locus. It can therefore be used for gene tracking in haemophilia A. Figure 2 also shows how the sister of the man with haemophilia has inherited the same band (X chromosome) from her mother as her affected brother and is therefore almost certainly a carrier. In this case, however, her X chromosomes each give a different band on the autoradiograph and so she is "informative" for prenatal diagnosis during the first trimester, should she want it.

DNA analysis in practice

Despite its convenient independence of knowledge of the mutation, several factors may limit the application of gene tracking. Firstly, the relevant family members must be available. For prenatal diagnosis of cystic fibrosis, when one or both of the mutations are unknown, a sample from a previously affected child is necessary to discover with which chromosome 7 restriction fragment length polymorphism pattern the cystic fibrosis mutation is coinheriting. Similarly, it may be important to have samples from a relative with Huntington's chorea or a boy with Duchenne muscular dystrophy. In X linked disorders, however, normal boys or the maternal grandfather may often allow the linkage phase with the DNA marker to be established.

Secondly, key family members must be heterozygous for a DNA sequence polymorphism—that is, the two homologous chromosomes have to be distinguishable. The prospect of achieving this in nearly all families has increased with the characterisation of new sets of highly variable sequences distributed throughout the genome, referred to above. Some initial family analysis is necessary, however, to establish which DNA marker is informative for that particular family. For this reason, referral before a pregnancy at risk is essential, because nothing can be promised until after the initial family study.

Thirdly, in addition to human errors common to other tests, such as switched or mislabelled samples, gene tracking has an inherent error rate due to recombination or crossing over. Considering any one of the 23 pairs of chromosomes, one or other parental chromosome is not transmitted to a child intact. Between each generation there is reciprocal exchange of varying sections of the chromosome pair so that, although the sperm or egg contains a single chromosome, it is actually a composite of the parental pair. This normal exchange between homologous chromosomes during gamete formation may lead to "separation" of the mutation from the restriction fragment length polymorphism pattern (allele) with which it had been coinheriting, and consequently a wrong genetic prediction. To date, with the exception of the Duchenne muscular dystrophy locus, which is huge and probably a recombination "hot spot," errors due to recombination occur in fewer than 1–2% of cases with the use of gene specific probes. When linked DNA probes are used, as in Huntington's chorea, the error rate in genetic prediction can be calculated from the observation of how consistently the DNA marker coinherited with the disease in studies of large families. Linked probes with a recombination frequency of more than 5% are not usually used. It is important to realise that the recombination frequency may not equate direct with the error rate, and often complicated calculations are required before the reliability of the test is explained to a particular family.

Finally, prediction by gene tracking may sometimes be critically dependent on the stated father being the biological father (especially in X linked disease as in fig 2). Unexpected paternity is one of the several clinical, ethical, or technical problems that may arise in applying these new techniques in clinical genetics.

Integrating genetics and medicine

Clearly, the ideal is to be able to define the exact mutation in each and every family, and there will be steady technical progress in this direction. But technical advances have to be matched with a much broader input of molecular genetics into all types of clinical practice, from the simplest practical procedure of banking a DNA sample from an affected individual or fetus, to the recognition that the investigation of a patient with a monogenic disease includes defining the mutation at the DNA sequence level. Furthermore, close collaboration between clinicians and scientists, and between clinical geneticists, obstetricians, and those caring for the families, has to be maintained free of professional, administrative, and financial barriers if clinical genetics is to play its part in helping couples restore family life and reproductive choice in the face of known genetic risks.

Production and use of therapeutic agents

N H Carey

In the past 30 or 40 years advances in cell biology, endocrinology, and immunology have led to the discovery of many molecules acting at long or short range on target cells. In addition to the well known protein hormones a series of protein growth and differentiation factors were discovered. These were identified as activities in tissue extracts or the products of in vitro cultures and were usually present in such small quantities that purification and structural studies were difficult or impossible. Even more remote was the opportunity to test them in pure form in disease models or clinically.

More recently advances in molecular biology have resulted in new techniques which permit these and other proteins to be made on a large scale. These techniques, collectively known as recombinant DNA technology, allow the transfer of genes in a predetermined way from one cell or organism to another. Thus, in principle, it becomes possible to make a protein in a cell that can be grown on a large scale, such as a micro-organism.

All proteins—whether they are hormones (such as insulin, growth hormone, or calcitonin), enzymes (such as the clot forming and clot lysis components of blood), or antibodies—have a structure that is determined by the gene that specifies them. Genes are composed of the chemical DNA. Proteins are linear molecules composed of many of the 20 amino acids; their structure is determined by the order in which the amino acids are arranged. This order is in turn determined by the order of the components of DNA, which is also a linear molecule.

Each cell has its component of DNA molecules, but not all of them are expressed—that is, used to give rise to the proteins they specify. The choice of genes that are expressed depends on the stage in the life cycle of the cell or on what kind of cell it is. Fibrinogen, for example, is made in the liver, haemoglobin in erythrocyte precursors, and myosin in muscles. The genes for these proteins are not expressed in cells that do not make them. This means that within each cell is a set of control mechanisms that determine which of its genes are expressed and in what order. Thus simply introducing a gene into a foreign cell does not guarantee that the protein product of that gene will be made by the cell. The correct control mechanisms are also needed.

For more than 15 years scientists have been able to manipulate genes in vitro with the aid of enzymes and some chemical synthetic procedures. The linear molecule of DNA can be cut, reassembled, and joined again to give new molecules with the genes recombined in a different arrangement. Genes from different organisms can be combined together, and the new collection of genes can be reintroduced into a cell where it is reproduced along with the cell's own genes. Because this process resembles the normal process of recombination that occurs during the reproductive cycle, this new form of DNA is called recombinant DNA.

If no other steps are taken when a recombinant DNA molecule is introduced into a cell it will simply be reproduced at every cell division. If, however, the gene is placed alongside the correct structures in the DNA that comprise the control mechanisms mentioned above then the gene can be expressed in its new cell, giving rise to the protein product that it specifies. The expression of recombinant DNA was first achieved with the bacterium *Escherichia coli*, an organism whose genetics have been so well studied that its requirements for expression were easiest to predict. Since then expression systems have been devised for many types of cell, including yeast and animal cells growing in culture. All of them allow for the controlled production of the protein in fermentation systems similar to those used for producing antibiotics. The cell system that a researcher chooses nowadays to express a protein depends on the nature of that protein and the use to which it will be put. For example, if a large quantity of the protein is required a micro-organism would probably be chosen as the means of production. If the protein is large and complex an animal cell is more likely to be chosen for culture.

One special technique for producing potential therapeutic agents is the monoclonal antibody technique, also referred to as the hybridoma technique. The end result of this approach is the production of a protein, the desired antibody, in a tissue culture cell; operationally, therefore, it is similar to the systems described above. The method of deriving the cell line is different, however, as it depends on the fusion of two different cell types to introduce the gene into the production cell rather than introducing it as a piece of purified DNA. The earlier methods of making antibodies—that is, injection of a suitable animal with the antigen—resulted in serum that contained a mixture of antibodies with a wide variety of properties. The hybridoma technique allows the selection of a cell making only one of the whole repertoire of antibodies that an intact animal would make when exposed to the antigen.

The processes of fermentation and protein purification that are used in the production of proteins from the cells engineered in the ways described above are well understood so that, although there have been some new problems and some special features to consider, the production of high quality therapeutic proteins is becoming increasingly easy.

Therapeutic use of recombinant protein

What are these proteins used for? What advantages do they bring, and what are their problems and limitations? The most important reasons for preparing therapeutic proteins in this way are that they cannot be obtained by more traditional means such as extraction from natural sources or, if they can, that the product is in some way unsatisfactory.

Human insulin

The first product of any consequence to be made by recombinant DNA techniques was human insulin. Diabetic patients have been treated with insulin extracted from animal pancreases (mainly pig and beef) for some time. Natural human insulin has not been available because, in the face of the satisfactory performance with the animal products, it was not considered worth while to set up the complex procedures for collecting human pancreases from cadavers, even if the supply would have been adequate. Now, however, a product identical to the natural human product is available through fermentation in *E coli*. Although there has been a suspicion that some patients respond differently to this form of insulin, there is an instinctive feeling that the human form of the hormone should be preferable to the animal versions. Most people agree, however, that any important benefits are likely to be long term.

Growth hormone

Unlike insulin, growth hormone is species specific. Patients who suffer from hypopituitarism and need replacement treatment must therefore be given the human hormone. Until recently this came from human pituitaries obtained from cadavers. Such treatment was considered satisfactory until suspicions were aroused that a few patients had been infected through the treatment with an agent that gave rise to the Creutzfeldt-Jakob syndrome, a human form of the spongiform encephalopathies. The natural product has now been largely removed from the market. Fortunately, replacements from several sources, made by recombinant DNA, were almost ready for introduction, and they have been developed rapidly.

This then is an example of the use of a recombinant product to substitute for a natural product which proved unsatisfactory because of contamination with an infectious agent. A similar problem has arisen with the contamination of factor VIII preparations, used for treating haemophilia, with HIV. In this case, however, the recombinant product is not yet available, apparently because of the great complexity of the protein. Alternative means have been used to make the factor VIII preparations safe.

Hepatitis vaccine

Hepatitis B is caused by a small DNA virus and has hitherto been impossible to control by a vaccine because the virus could not be grown in any of the normal laboratory systems used for preparing viruses—that is, tissue culture cells or embryonated eggs. Until recently the most promising vaccine has been a killed virus preparation made from the plasma of infected individuals. This has obvious drawbacks. The collection process is laborious, and there is no possibility of developing safer attenuated forms of the virus as has been done, for example, for polio. Ultimately, the more successful the vaccine the more difficult it would be to prepare because the number of infected individuals would decrease. It has to be admitted, however, that this limitation is a long way off.

Again, the solution to this problem has been to clone the appropriate gene from the virus into a micro-organism and to express it to obtain viral antigen. One of the more promising preparations has come from a yeast clone of this gene. This preparation is safe both for the patient and for the people responsible for its manufacture because there is no virus present. The yeast cell contains the information for only one gene from the virus. Thus there is no way that a complete virus or any other infectious agent could arise in the vaccine.

This technology therefore represents a promising way of making vaccines for agents that are difficult or dangerous to grow in culture or where the vaccine itself may be a danger to the patient, who is not usually ill at the time of treatment. It will undoubtedly be central to the production of a vaccine for AIDS. The techniques are also being used to dissect the molecular biological intricacies of the agents causing malaria, schistosomiasis, and trypanosomiasis, with the eventual aim of control by vaccine.

Anticancer monoclonal antibodies

The immune system has long been implicated in the body's defence against cancer, a defence that obviously goes awry when the disease becomes established. This thought has led to the proposition that stimulating or substituting for the presumed defects of the endogenous system may be a route to controlling or curing the disease.

One of the routes of attack has been to try to obtain tumour specific antibodies and to attach chemotherapeutic or other cell killing agents to them. If successful this would result in the targeting of the killing agent on the tumour, greatly reducing the toxicity in the rest of the body. Promising results were obtained some time ago with polyclonal antibodies in animal models, but for various reasons these could not be reproduced in man. One of the reasons may lie in the fact that a polyclonal serum contains many different antibodies and, indeed, many non-antibody proteins. If an antibody to a particular tumour cell determinant was required, its concentration in the mixture would be so low that it would be ineffective. Monoclonal antibodies clearly represent a way of overcoming this problem. High concentrations of a single antibody that is specific to a selected determinant on the target cell can be obtained.

Much work has been done to discover appropriate antibodies, and many promising candidates have been found. One important principle has been agreed on as a result of this work—that is, that it is unlikely that a truly tumour specific antigen exists. "New" antigens that appear on tumours have been found on other cells at some time in the lifetime of the individual, either during development or on stem cells in the tissue in which the tumour originates. This means that if antibodies can be used as cytotoxic agents care will have to be taken over dosage and distribution among tissues and cells.

One further feature of antibodies that needs to be taken into account is that they are not designed to be used as antitumour agents. There are therefore several features of the antibody molecule that could militate against its usefulness. The methodology of recombinant DNA now provides the opportunity of modifying the structure of antibodies in a controlled way to overcome this problem. The genes for a candidate antibody may be cloned, then modified to alter or delete unwanted portions, and finally reinserted into a suitable cell for production.

It is too early to say how successful this approach will be in a wide variety of tumours. The early results are promising, however, and it seems likely that modified antibodies will soon be among the collection of drugs used to control tumour growth.

Limitations of the technology

Many other important products under development are emerging from the use of this technology in health care, and some are already licenced for clinical use in several countries. Examples are tissue plasminogen activator for use in myocardial infarction, erythropoietin for use in various forms of anaemia, and lymphokines and growth factors, which may find a variety of uses.

The main limitation of this approach is that all the therapeutic agents are proteins. They cannot be taken by mouth because they would not survive exposure to the enzymes in the gut and would not be absorbed from it. They must therefore be given by injection, which limits their use mainly to patients in hospital. Scientists in many organisations, both academic and commercial, are working on approaches that may overcome this problem. These involve, for example, slow release depot formulations or compositions that protect the protein from enzyme attack or assist its penetration through mucous membranes and into the blood stream. These methods are meeting with some success, but it still seems likely that most of these products will need to be given by injection for some time to come.

Treatment of the diseases under consideration by the first wave of products coming from the technology will not be hindered by this problem, as the patients are already being treated in hospitals or clinics. There are many other conditions, however, for which frequent attendance at hospital and treatment by injection is undesirable or impossible. These include the various forms of arthritis, hypertension, and osteoporosis. This does not mean that molecular biology has nothing to contribute to the discovery of treatments for these diseases.

New approaches for the future

The development of modern drugs in the past 30 years or so has depended on our understanding of the basic biology of the disease and the quality or relevance of the model systems used to test potential drugs. It has become apparent during this period that active compounds achieve their effects by combining with macromolecules in the organism, termed receptors, and consequently modifying or mimicking the effect of a natural molecule in vivo.

Many drugs have been discovered and developed without knowledge of the nature or function of the receptor concerned. It is clear, however, that a better understanding of the nature of the receptor and its natural activation will help the design of active drugs. This is where molecular biology will have an important impact on drug design. The genes for many receptors and their activators are now being cloned. In many cases this will lead to the design of model systems using the receptor, which would not be possible in the natural state. In the longer term the structures of the natural molecules can be elucidated by *x* ray crystallography or by other techniques, which will greatly facilitate the study of their interaction using computer graphic modelling. This in turn will permit a much more precise design of potentially active compounds.

The polymerase chain reaction: a tool for molecular medicine

A F Markham

The polymerase chain reaction has been unquestionably unique, as new techniques go, in the speed with which it has been embraced by non-experts in most specialties of the biological sciences, including medicine. The reason for this is the unusual simplicity of the procedure. In terms of its power to drive biological research, the advent of the polymerase chain reaction can certainly be compared with the discovery of the techniques of molecular cloning some 20 years ago. However, whereas years of training and practice were usually needed to master the many and complex skills of recombinant DNA technology, the complete beginner can start to perform polymerase chain reaction experiments and generate meaningful results within a few days at most—hence the explosion of activity.[1–1c]

The technique was first described in its initial format in 1985,[2] and over the next three years appreciation of its potential gradually became widespread. This potential was fully realised in about 1988. It coincided with the commercial development of two key components for the polymerase chain reaction: a DNA polymerase that could be heated at quite high temperatures (boiling water) without losing its activity, and robust machines that would quickly heat and cool samples repeatedly in a cyclic fashion.[3] Synthesis of the oligonucleotides required as primers in the reaction had already become a commonplace procedure. In the past four years the reaction has become probably the most widely used single technique in all branches of the biological sciences.

What is the polymerase chain reaction?

The polymerase chain reaction is a delightfully simple concept, first alluded to 30 years ago,[4] and very reminiscent of the way that cells duplicate their DNA to expand their numbers in vivo. The other simple analogy is to the chain reaction of nuclear physics. The technique permits the analysis of nucleic acids (DNA or RNA) from any source.

Procedure

A small sample of DNA in solution is placed in a single tube. (As will become clear below, any RNA samples for analysis are simply converted to DNA in a single preliminary step.) Two oligonucleotides (which are easy to make artificially on automatic machines or can be purchased from several suppliers) are added to the tube. Their sequences are chosen so that they match two short DNA sequences that flank the region of interest. The exponential increase in the amount of DNA sequence that is now known, and the universal accessibility of this data through computer databases, mean that any scientist can quickly start to study genes of interest in his or her own laboratory. There is no need to obtain DNA clones from other workers, so all the problems previously associated with that are avoided.

Considerably more oligonucleotides than the DNA to be analysed are provided deliberately. A thermostable DNA polymerase is added to the same tube.[5] Deoxynucleoside triphosphates (dNTPs) and salts and buffer to allow the enzyme to work properly are also included. All these reagents are available commercially. Taq DNA polymerase was isolated from the thermophilic organism, *Thermus aquaticus*. Not only can it tolerate heating at 100°C, but it makes DNA at high temperatures compared with the 37°C physiological temperature optimum of most enzymes. A cloned version of this enzyme without any nuclease activity is also available.

This simple mixture is placed in a heater and the temperature quickly raised to just below the boiling point of water. This causes the double stranded DNA in the sample to dissociate into two single strands as the hydrogen bonds, which hold the two strands together under physiological conditions, break down reversibly on heating. After a minute or so, the solution is allowed to cool towards physiological temperature. This allows hydrogen bonds to re-form. The two DNA strands in the sample would, of course,

Outline of the procedure

- A small sample of DNA is placed in a tube
- Two oligonucleotides are added. These have sequences matching two sequences of the DNA that flank the region of interest
- A thermostable DNA polymerase is added
- The mixture is heated to just below 100°C and the DNA dissociates into two single strands
- The solution is allowed to cool and the single strands bind to the oligonucleotides, which are in excess
- The oligonucleotide now acts as a primer for DNA polymerase and is extended to form a new double stranded molecule
- The cycle is repeated, with the amount of DNA doubling each time

usually relocate their partners and re-form the paired double helix. However, in the polymerase chain reaction tube the oligonucleotides, which are present in great excess, quickly and highly specifically bind to their complementary single strands from the denatured sample. As soon as this happens the oligonucleotide can act as a primer for DNA polymerase and is extended to form a new double stranded molecule. Thus each double stranded DNA molecule in the original sample has been melted to form to single stranded molecules, which have then been turned into two double stranded molecules.

There is now twice as much double stranded sample DNA present in the tube as there was to start with. The cycle is then repeated and the amount of sample DNA doubled further with every cycle. This geometrical amplification is perfectly analogous to a nuclear chain reaction with 2-fold, 4-fold, 8-fold, 16-fold, 32-fold, etc amplification at subsequent steps. After n cycles the degree of amplification is of course 2^n. Thus after 10 cycles 1024-fold amplification is achieved and after 20 cycles 10^6-fold amplification results.

No great manual dexterity is required to perform the technique, in that all these reactions go on by simply automatically heating and cooling without ever opening the reaction tube. The whole process is outlined in figure 1.

The orientation of the two strands in double stranded DNA is important to note. They are described as being antiparallel, which means that when bound together in a helix one reads 5′ to 3′, the other 3′ to 5′. This simple consideration dictates the design of the

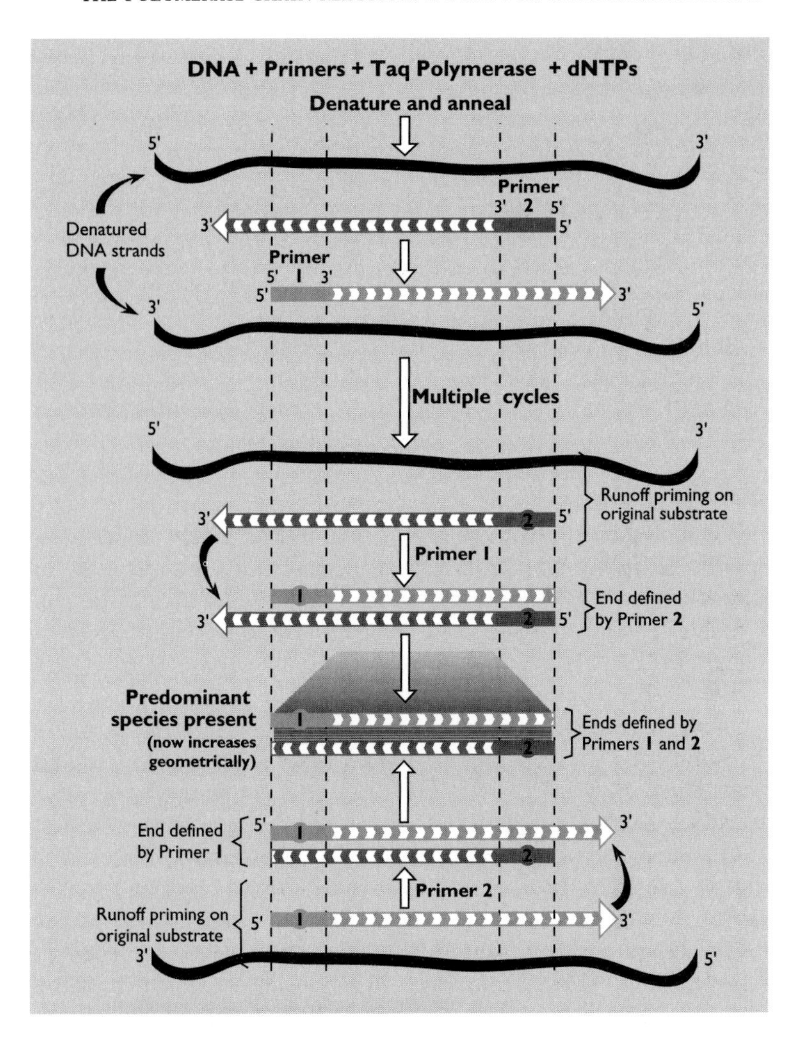

FIG 1—Schematic illustration of the polymerase chain reaction. Initially runoff priming occurs. In subsequent cycles the product with ends defined by both Primers 1 and 2 becomes predominant. It increases geometrically in amount by doubling with each cycle. After 30 cycles there are a billion copies of the reaction product but only 30 copies of the runoff products.

synthetic oligonucleotides. When making double stranded DNA, Taq polymerase attaches new residues to the 3′ end of primers. Thus in the polymerase chain reaction DNA is essentially copied only between the two primers. After multiple cycles of amplification, the predominant double stranded DNA species in the sample is a fragment whose two ends are defined by the two primer oligonucleotides—that is, not all the DNA in the original sample is amplified. For example, if the sample DNA were total human DNA from a blood sample, and therefore contained 3×10^9 base pairs, amplification of any 300 base pair segment could be specifically achieved. It will be appreciated that the signal to noise problems inherent in detecting and analysing 300 base pairs out of 3 billion (1 in 10^7) are very much alleviated after 30 cycles of polymerase chain reaction, when 10^9 copies of the 300 base pair fragment are present for every single copy of the total genomic DNA background.

Problems

The polymerase chain reaction is not quite infinitely powerful or infallible. Eventually, after many cycles, the concentration of oligonucleotide primers falls because they have all been incorporated into products of the reaction. Similarly, the concentration of deoxynucleoside triphosphate substrates, and indeed the activity of the DNA polymerase itself, decline. Products of pyrophosphate breakdown may inhibit further reaction. Amplification will usually be sufficient for most analytical purposes by this stage, but should even greater sensitivity be necessary—for example, in the analysis of DNA from a single cell—then a tiny aliquot from the first polymerase chain reaction tube (as a source of sample DNA) is simply transferred to a second identical tube, and amplification is continued. Should specificity be a problem (for instance, if the target DNA comes in a sample that also contains many other closely related DNA sequences) then this can be overcome using "nested" primers in the second tube. These match sequences just inside the two original primer sites. Thus nested amplification of a product of the polymerase chain reaction depends overall on accurate recognition by four independent oligonucleotides.

The other aspect that may cause problems is the specificity of the priming reaction itself. This depends on several considerations. The size of the oligonucleotide will determine whether it occurs more than once in a sample DNA and therefore might prime DNA polymerase activity at multiple sites. This will generate only spurious products of the polymerase chain reaction when unwanted priming also occurs close by on the other DNA strand. However, the reaction may be inefficient if one of the primers is depleted because of excessive spurious priming. Several simple measures can be taken to eliminate this type of problem should it arise. These include using raised temperatures during cycling, adjusting the magnesium concentration (which partly determines the ease of hybridisation), and various other easy tricks to destabilise partially mismatched primers. On balance, these considerations do not usually constitute a serious problem, and any teething troubles can usually be overcome quickly by minor and obvious adjustment to reaction conditions.

Why has the reaction had such a major impact?

Southern blotting

The impact of the polymerase chain reaction can be attributed to its practical simplicity and to its speed, sensitivity, and specificity. Consideration of some of the routine day to day techniques of molecular biology allows this to be illustrated. In a genomic Southern blot it is impractical to load more than 10 μg (10^{-6} g) of restriction enzyme digested DNA per lane of an agarose gel. Higher loading causes overloading and streaking. If one is interested in a 3 kilobase fragment then this constitutes about 10 pg (10^{-12} g) of the sample. Detecting 10 pg of anything is taxing, and doing so when it is effectively "contaminated" with a million-fold excess of somewhat similar DNA fragments is even harder. The practical consequence has previously been that prolonged autoradiographic techniques were required. Days or weeks would be needed for each experiment. By simply amplifying the fragment of interest it can be characterised in agarose gels, usually by visual inspection, in a few hours.

Messenger RNA

Another example of the reaction's impact would be the use of the polymerase chain reaction to overcome some of the many problems inherent in analysing messenger RNA (mRNA) by classic techniques. These molecules, the ultimate source of information about what is going on at a specific time in a particular cell, are extremely labile chemically (for example, to traces of alkaline detergent in less than scrupulously clean glassware) and enzymatically (to the ubiquitous ribonuclease). A typical RNA preparation from a tissue sample or cell culture will contain around 2% of mRNA or less, with 98% ribosomal RNA (rRNA) and transfer RNA (tRNA). Though selecting for the polyadenylated mRNA on oligo(dT) affinity columns is possible, it is notoriously difficult even for the nimble fingered. Handling the minute amounts of material required remains extraordinarily difficult. Synthesis and subsequent cloning of copy DNA (cDNA) often yielded unsatisfactory clone banks with short inserts and heavy ribosomal DNA (rDNA) contamination for these purely technical reasons.

The situation is transformed by the use of the RNA polymerase chain reaction (fig 2). Oligo(dT) itself can be used to prime first strand synthesis of cDNA. Commercial kits are available and even

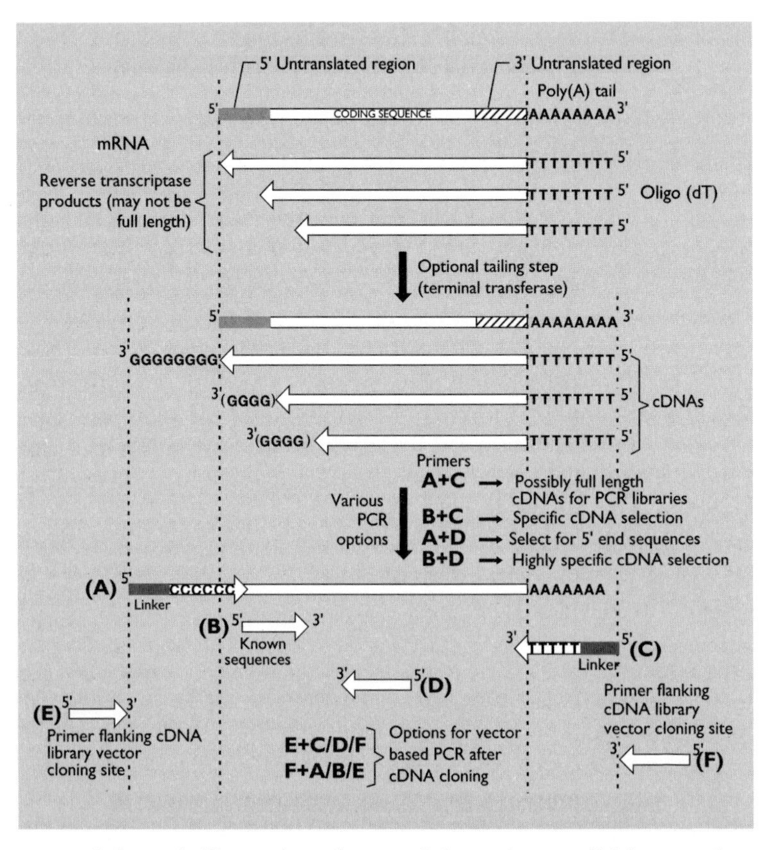

FIG 2—Schematic illustration of some of the options available to analyse mRNA by the reverse transcriptase polymerase chain reaction. When single stranded cDNA has been synthesised a variety of products can be generated by selecting pairs of the primer types A to F. As illustrated for primers A and C, linker sequences may be included at the 5′ ends of any of these primers to further facilitate cloning of the resulting reaction products.

thermostable "reverse transcriptases" have been introduced. Double stranded cDNA is obtained by a number of methods including using a second oligonucleotide[6] or degenerate mixture specific for a target mRNA, or else total double stranded cDNA is produced by "tailing" the first strand and using the complementary homo(oligonucleotide) as the second primer.[7] This has become known as "anchor polymerase chain reaction." After multiple cycles of amplification contaminating RNA species are no longer a problem and there is plenty of material to clone, sequence, or otherwise analyse, all within a few hours and with minimal manual intervention.

Incorporating the reaction into routine molecular genetics

Analysis of restriction fragment length polymorphisms

Analysis of restriction fragment length polymorphisms, previously a stock-in-trade of molecular genetics, has been revolutionised by the polymerase chain reaction. We have discussed the inherent difficulty of Southern blotting above. For restriction fragment length polymorphisms analysis it is essential, in particular, to generate DNA of sufficient quality to undergo digestion to completion with restriction enzymes. This is not always easy. Failure to cut sample DNA may lead to diagnositc errors with clinical material. Amplifying the target DNA means that obtaining high quality DNA should never be a problem, at least in principle. Control amplifications on samples to show that a constant restriction enzyme site can be cut strengthen confidence in the analysis. Furthermore, the presence or absence of polymorphisms can be assessed immediately by agarose gel electrophoresis and by inspection of ethidium bromide stained products under ultraviolet light.

The reaction has in fact permitted dramatic advances in the use of genetic polymorphism in linkage analysis. As well as classic dimorphic restriction fragment length polymorphisms, which are not particularly informative, several classes of repeat sequences have now been characterised in DNA which are highly polymorphic as a consequence of wide variation in repeat copy number.[8] This is easily assessed by the polymerase chain reaction using sequences that flank the microsatellite or minisatellite repeat—the size of the product reflects the number of repeat units. The

Uses of the reaction in molecular genetics

- Analysis of restriction fragment length polymorphisms
- Analysis of messenger RNA
- Amplification of fragments for identification by Southern blotting
- Assessment of genetic polymorphisms in linkage analysis
- DNA sequencing
- Preparation of elusive DNA fragments for cloning

procedure has revolutionised human genetics by massively increasing the speed and power of pedigree analysis. Of importance to the clinician, certain inherited diseases (myotonic dystrophy, the fragile X syndrome, Kennedy disease) seem to be the result of spontaneous increases in the copy number of trinucleotide microsatellite repeats.[9] Again, assays for this based on the reaction already permit presymptomatic or antenatal diagnosis.

DNA sequencing

DNA sequencing, another fundamental molecular genetic technique and increasingly a vital component for identifying molecular pathology in patients, has also become reliant on the reaction. Traditionally, single stranded DNA for sequencing was obtained by cloning into the M13 bacteriophage system and subsequently purifying phage DNA. Given a reasonable amount of DNA template, sequencing was achieved by extending oligonucleotide primers in the presence of dideoxynucleoside triphosphate chain terminators. Amplification by the reaction solves the DNA yield problem without cloning. The material may be either sequenced double stranded or rendered single stranded by asymmetrical polymerase chain reaction (using very much less of one of the two primers),[10] selective enzymatic digestion of one of the strands, or by selective capture of one of the strands by labelling one oligonucleotide with biotin and immobilising the product on streptavidin coated magnetic beads. So called "cycle sequencing" can increase the sensitivity of the process. In many circumstances the use of high temperatures for the thermostable DNA polymerase reaction is advantageous in sequencing because secondary structure in the template strand is eliminated.

Contamination

There can be problems in controlling this powerful technique. As the polymerase chain reaction is able to amplify even a single molecule, contamination by any previously amplified DNA would be catastrophic. Many precautions have been devised to avoid contamination,[11] but scrupulous housekeeping is essential in any laboratory routinely undertaking the reaction. An extensive debate about the fidelity of copying DNA sequences by Taq polymerase has been published, but this is not usually an issue except in highly specialised applications, such as the study of variant clinical isolates of HIV.[12]

Some applications of the reaction

There is an enormous range of novel applications of the reaction, and the list of references contains several comprehensive reviews. Specific examples that have been particularly useful include:

Quantitative polymerase chain reaction, which permits measurement of the level of specific mRNAs in different cell populations.[13] The amount of target product of the reaction is compared with the amount generated from a control amplification target in the same reaction. The technique has been most widely applied in the analysis of cytokine responses.

RACE-polymerase chain reaction. RACE stands for rapid amplification of cDNA ends. The reaction is performed between one primer designed on the basis of a given protein sequence and a second, which flanks the cloning site in a phage, or plasmid vector. The substrate is a total cDNA library. The product of the reaction is derived from the targe mRNA, and the technique is particularly useful for identifying 5′-terminal cDNA sequences.

Alu-polymerase chain reaction is enormously valuable.[14] Reaction primers match sequences in the ubiquitous interspersed repetitive sequences, of which about a million copies are scattered throughout the human genome. Amplification yields reaction products derived from the regions between Alu sequences. As the Alu repeats are relatively unique to humans, the human component from sources such as somatic cell hybrids in rodent cell backgrounds, or human DNA cloned in cosmids or yeast artificial chromosomes (YACs), can be isolated.

Inverse polymerase chain reaction permits amplification of DNA in which the sequence of only one end of the fragment is known. This permits effective chromosome walking.[15 16]

Vectorette polymerase chain reaction also permits amplification of regions of DNA of unknown sequence that flank known sequences. The approach entails restriction digestion of the DNA sample then ligation of specially designed "vectorette" linkers. The amplification is performed between the known sequence and the vectorette sequence. This method has proved generally useful in chromosome walking exercises and is particularly convenient for isolating the ends of yeast artificial chromosome clones for rapid physical genetic mapping and ordering of overlapping YACs to assemble so called "contigs".[17]

Specific applications of the reaction in molecular medicine

There are three distinct types of clinical challenge for which the polymerase chain reaction is indispensable: (1) Detecting vanishingly small amounts of DNA so as not to miss even the most cryptic infection and permit analysis of single cells or single sperm and of partially degraded samples. (2) Identifying the specific new mutation in a particular gene that causes a given inherited disease in a patient. This is necessary to gain an understanding of the molecular basis of the resulting disease and also to allow accurate family studies for genetic counselling. (3) Analysis to detect known

Present uses of the reaction in medicine

- Detection of vanishingly small amounts of nucleic acid—for example, in HIV infection
- Identification of new genetic mutations
- Routine detection of known mutations—for example, Duchenne type muscular dystrophy, cystic fibrosis assays based on the reaction allow presymptomatic or antenatal diagnosis of diseases due to microsatellite instability such as myotonic dystrophy, the fragile X syndrome, and Kennedy disease
- Detection of mutations thought to predispose to disease—for example, myocardial infarction and cancer
- Detection of mutations in malignant tumours to assess prognosis
- HLA subtyping

mutations that always cause a particular inherited disease or polymorphism (for example, sickle cell anaemia, or the cystic fibrosis F508 deletion).

Detecting vanishingly small quantities of nucleic acid has been achieved in the diagnosis of HIV infection and in such unusual applications as detecting measles virus RNA in brain biopsy specimens from patients with subacute sclerosing panencephalitis. The nested technique is often helpful, and eventually the products of polymerase chain reactions may even be analysed by Southern blotting to gain further absolute sensitivity. In these circumstances, the issue of sample contamination giving false positive results is usually as important as the risk of false negative results arising from "failure" of the reaction. The ability to detect negligible (a few molecules) DNA is also essential for preimplantation diagnosis in in vitro fertilisation clinics.[18]

Characterisation of gene mutations

A range of elegant techniques has been developed for characterising new genetic mutations—all the techniques rely on the generation by the polymerase chain reaction of large quantities of DNA fragments from normal and mutant alleles. The problem is to detect as little as a single base pair difference between, say, two 500 base pair fragments. This is often achieved by creating heteroduplexes between the two alleles and looking for modified properties in: (*a*) chemical cleavage reactions; (*b*) ribonuclease digestion (one fragment is RNA); (*c*) denaturing gradient gel electrophoresis; (*d*) single strand conformational polymorphism; (*e*) reaction with carbodiimide; or (*f*) automated total sequencing of both fragments.

A smaller number of techniques based on the polymerase chain reaction is available for the routine detection of known mutations. If the mutation creates or destroys a restriction site then this can be simply examined in the products of the reaction. The disadvantage is that two distinct operations (polymerase chain reaction and restriction cleavage) are required. Detecting mutations by hybridising products of the reaction with allele specific oligonucleotides is possible, although again the need for two distinct procedures is a drawback.

The concept of an allele specific polymerase chain reaction, usually called "ARMS" (the amplification refractory mutation system), is rather more convenient and general.[19] Here one of the pair of reaction primers is deliberately designed so that its 3′

terminal residue lies precisely at the point mutation site. Two experiments are run in parallel: in one the allele specific primer matches the normal sequence, in the other it matches the mutant sequence. A product is obtained from the reaction only when 3′ residues form base pairs and prime synthesis correctly. By comparing the products in the "normal" and "mutant" reactions, normal and mutant homozygotes as well as heterozygote carriers are easily identified. Many genetic diseases are now routinely tested for in this way.

Furthermore, several pairs of the reaction primers can be mixed in the same tube and allowed to "analyse" different point mutations at the same time (all the different products of the polymerase chain reaction are deliberately designed to be different sizes). By this "multiplex" approach, for example, all of the four (or more) common mutations causing cystic fibrosis can be looked for simultaneously in a single sample.[20] Several similar assays based on thermostable DNA ligases rather than polymerase are also being developed. A similar type of multiplex polymerase chain reaction analysis aimed at detecting DNA deletions has revolutionised antenatal diagnosis of Duchenne type muscular dystrophy. Another clinical advantage of these exquisitely sensitive tests has been that sufficient DNA can be obtained from buccal washings, dried blood on Guthrie phenylketonuria test cards, or tiny chorionic villus biopsy specimens.

Applications

As the technique has developed so the range of applications in clinical practice has expanded. HLA molecular subtyping by polymerase chain reaction is straightforward, and clinicians are now attempting to develop tests that predict people at risk of developing insulin dependent diabetes mellitus on that basis.[21] Mutations that appear to predispose to myocardial infarction (in the angiotensin converting enzyme (ACE) gene)[22] and hypertension (in the angiotensinogen gene)[23] have been identified recently. Population screening programmes may be worth while. Both predispositions may be suppressible by treatment with angiotensin converting enzyme inhibitors. Various mutations in both oncogenes and tumour suppressor genes have been implicated heavily in the development of cancer. Assessment of the extent of such mutation in a given tumour may permit assessment of prognosis or predict whether distant metastasis has already occurred.

Future clinical possibilities

This raft of techniques can be applied to most specialties. Here I will limit discussion to two disciplines in which there is great activity.

Antenatal diagnosis

In antenatal diagnosis much effort is directed towards methods of analysing those few fetal cells found circulating in a pregnant woman's peripheral blood. By this approach invasive fetal sampling techniques may be avoided entirely or limited to those cases where confirmation of a positive diagnosis is desirable.

The estimate is that less than 10 000 fetal nucleated erythrocytes are present in 20 ml of maternal blood. Fluorescence activated cell sorting with anti-CD71, anti-CD36, or anti-glycophorin A, or a combination of these three monoclonal antibodies, permits enrichment towards 90% pure fetal nucleated erythrocytes. Trisomies can be detected direct by fluorescence in situ hybridisation. It seems to be only a matter of time before sufficient purification is achieved routinely to permit the full range of diagnosis based on the polymerase chain reaction as above. Indeed, this whole topic was resurrected some three years ago by the demonstrations of Y chromosome specific product obtained by the technique from pregnant women with male fetuses.[24] This essentially confirmed that fetal tissue was in fact present in the maternal circulation.

Detection of cancer

Early detection of cancers is widely regarded as clinically important in that early treatment usually improves prognosis, and curative surgery may even be possible—for example, in colorectal cancer. The technique will have a major impact in this subject. It is already clear that mutations in p53 genes indicative of developing bladder cancer can be detected by analysis of shed cells present in patients' urine samples.[25] Even more remarkably, premalignant changes in the gastro-intestinal tract can be detected by testing faeces using the polymerase chain reaction. Enough premalignant cells are present in the bulk of stool to permit the analysis of tumour suppressor gene mutations by this technique. Patients at high risk can thus be selected for colonoscopy. Whether this approach offers advantages over faecal occult blood testing remains to be established.[26,27] The fact that gastric *Helicobacter pylori* infection has been assessed from stool samples by this technique suggests that gastric malignancy may also eventually prove detectable in this way. Detection of potentially metastatic cells in the circulation of patients with newly diagnosed primary tumours (for example, melanoma) is particularly intriguing.[28] The impending characterisation of a melanoma related tumour suppressor gene from human chromosome 9 should make this type of approach even more useful.[29] The clinical usefulness of the polymerase chain reaction thus seems to be limited only by the power of our imagination in identifying specific targets.

1 White TJ, Arnheim N, Erlich HA. The polymerase chain reaction. *Trends in Genetics* 1989;**5**:185–9.
1a Erlich HA, ed. *PCR technology*. New York: Stockton Press, 1989.
1b Innis MA, Gelfand DH, Sninsky JJ, White TJ, eds. *PCR protocols*. New York: Academic Press, 1990.
1c McPherson MJ, Quirke P, Taylor GR, eds. *PCR, a practical approach*. Oxford: Oxford University Press, 1991.
2 Saiki RK, Scharf S, Faloona F, Mullis KB, Horn GT, Erlich HA. Enzymatic amplification of β-globin genomic sequences and restriction site analysis for diagnosis of sickle cell anemia. *Science* 1985;**230**:1350–4.
3 Saiki RK, Gelfand DH, Soffel S, Scharf SJ, Higuchi R, Horn GT, *et al.* Primer-directed enzymatic amplification of DNA with a thermostable DNA polymerase. *Science* 1988;**239**:487–9.
4 Kleppe K, Ohtsuka E, Kleppe R, Molineux L, Khorana HG. Studies on polynucleotides XCVI: repair replications of short synthetic DNAs as catalysed by DNA polymerases. *J Mol Biol* 1971;**56**:341.
5 Chien A, Edgar DB, Trela JM. Deoxyribonucleic acid polymerase from the extreme thermophile Thermus aquaticus. *J Bacteriol* 1976;**127**:1550.
6 Frohman MA, Dush MK, Martin GR. Rapid production of full-length cDNAs from rare transcripts: amplification using a single gene-specific oligonucleotide primer. *Proc Natl Acid Sci USA* 1988;**85**:8998–9002.
7 Loh EY, Elliott JF, Cwirla S, Lanier LL, Davis MM. Polymerase chain reaction with single-sided specificity: analysis of T cell receptor delta chain. *Science* 1989;**243**:217–20.
8 Weber JL, May PE. Abundant class of human DNA polymorphisms which can be typed using the polymerase chain reaction. *Am J Hum Genet* 1989;**44**:388–96.
9 Richards RI, Sutherland GR. Dynamic mutations: a new class of mutations causing human disease. *Cell* 1992;**70**:709–12.
10 Gyllesten UB, Erlich HA. Generation of single-stranded DNA by the polymerase chain reaction and its application to direct sequencing of the HLA-DQA locus. *Proc Natl Acad Sci USA* 1988;**85**:7652–6.
11 Sarkar G, Sommer SS. Shedding light on PCR contamination. *Nature* 1990;**343**:27.
12 Tindall KR, Kunkel TA. Fidelity of DNA synthesis by the Thermos aquaticus DNA polymerase. *Biochemistry* 1988;**27**:6008–13.
13 Chelly J, Kaplan JC, Maire P, Gautron S, Kahn A. Transcription of the dystrophin gene in human muscle and non-muscle tissue. *Nature* 1988;**333**:858.
14 Nelson DL, Ledbetter SA, Corbo L, Victoria MF, Ramirez-Solis R, Webster TD, *et al.* Alu polymerase chain reaction: a method for rapid isolation of human-specific sequences from complex DNA sources. *Proc Natl Acad Sci USA* 1989;**86**:6686–9.

15 Triglia T, Peterson MG, Kemp DJ. A procedure for in vitro amplification of DNA segments that lie outside the boundaries of known sequences. *Nucleic Acids Research* 1988;**16**:8186.
16 Ochman H, Gerber AS, Hartl DL. Genetic applications of an inverse polymerase chain reaction. *Genetics* 1988;**120**:621–3.
17 Riley JH, Butler R, Ogilvie D, Finniear R, Jenner DE, Powell S, *et al.* A novel rapid method for the isolation of terminal sequences from yeast artificial chromosome (YAC) clones. *Nucleic Acids Research* 1990;**18**:2887–90.
18 Handyside AH, Pattinson JK, Penketh RJA, Delhanty JDA, Winston RML, Tuddenham EGD. Biopsy of human preimplantation embryos and sexing by DNA amplification. *Lancet* 1989;**i**:347–9.
19 Newton CR, Graham A, Heptinstall LE, Powell SJ, Summers C, Kalsheker N, *et al.* Analysis of any point mutation in DNA. The amplification refractory mutation system (ARMS). *Nucleic Acids Research* 1989;**17**:2503–16.
20 Ferrie RM, Schwarz MJ, Robertson NH, Vandin S, Super M, Malone G, *et al.* Development, multiplexing and application of ARMS tests for common mutations in the CFTR gene. *Am J Hum Genet* 1992;**51**:251–62.
21 Todd JA, Bell JI, McDevitt HO. HLA-DQ beta gene contributes to susceptibility and resistance to insulin-dependent diabetes mellitus. *Nature* 1987;**329**:599–604.
22 Cambien F, Poirier O, Lecerf L, Evans A, Cambou JP, Arveiler D, *et al.* Deletion polymorphism in the gene for angiotensin-converting enzyme is a potent risk factor for myocardial infarction. *Nature* 1992;**359**:641–4.
23 Jeunemaitre X, Soubrier F, Kotelevtsev YV, Lifton RP, Williams CS, Charru A, *et al.* Molecular basis of human hypertension: role of angiotensinogen. *Cell* 1992;**71**:169–80.
24 Lo Y-MD, Patel P, Wainscoat JS, Sampietro M, Gillmer MD, Fleming KA. Prenatal sex determination by DNA amplification from maternal peripheral blood. *Lancet* 1989;**ii**:1363–5.
25 Sidransky D, Von Eschenbach A, Tsai YC, Jones P, Summerhays I, Marshall F, *et al.* Identification of p53 gene mutations in bladder cancers and urine samples. *Science* 1991;**252**:706–9.
26 Powell SM, Zilz N, Beazer-Barclay Y, Bryan TM, Hamilton SR, Thibodeau SN, *et al.* APC mutations occur early during colorectal tumorgenesis. *Nature* 1992;**359**:235–7.
27 Sidransky D, Tokino T, Hamilton SR, Kinzler K, Levin B, Frost P, *et al.* Identification of *ras* oncogene mutations in the stool of patients with curable colorectal tumors. *Science* 1992;**256**:102–5.
28 Smith B, Selby P, Southgate J, Pittman K, Bradley C, Blair GE. Detection of melanoma cells in peripheral blood by means of reverse transcriptase and polymerase chain reaction. *Lancet* 1991;**ii**:1227–9.
29 Fountain JW, Graw SL, Kao W, Stanton VP, Aburatani H, Munroe DJ, *et al.* Further characterisation of the 9p21 region frequently deleted in human cutaneous melanoma. *Am J Hum Genet* 1992;**51**:A51.

Gene regulation

David S Latchman

That the expression of human genes must be a highly regulated process should be clear to anyone who has ever dissected a human body. The range of different tissues and organs is vast, they differ dramatically from each other, and they all synthesise different proteins—haemoglobin in red blood cells, myosin in muscle, albumin in the liver, and so on. Moreover, with few exceptions all these different cell types contain the same sequence of DNA, which encodes all these different cell proteins, and this DNA is also identical to the DNA in the single celled zygote, from which all these different cells arise during embryonic development. Clearly, therefore, some process of gene regulation must operate to decide which genes within the DNA will be active in producing proteins in each cell type.

Levels of gene regulation

Several stages exist between the DNA itself and the production of a particular protein (fig 1).[1] Thus the DNA must first be transcribed into a primary RNA transcript, which is subsequently modified at both ends by the addition of a 5′ cap and a 3′ tail of adenosine residues. Moreover, within this primary transcript, the RNA sequences which actually encode the protein are not present as one continuous block. Rather they are broken up into segments (exons) which are separated by intervening sequences (introns) that do not contain any protein coding information. As these introns interrupt the protein coding region and would prevent the production of an intact protein they must be removed by the process of RNA splicing[2] before the mature messenger RNA (mRNA) can be transported from the nucleus to the cytoplasm and translated into protein.

Clearly each of these stages is a potential point at which gene expression could be regulated, and there is evidence that several of

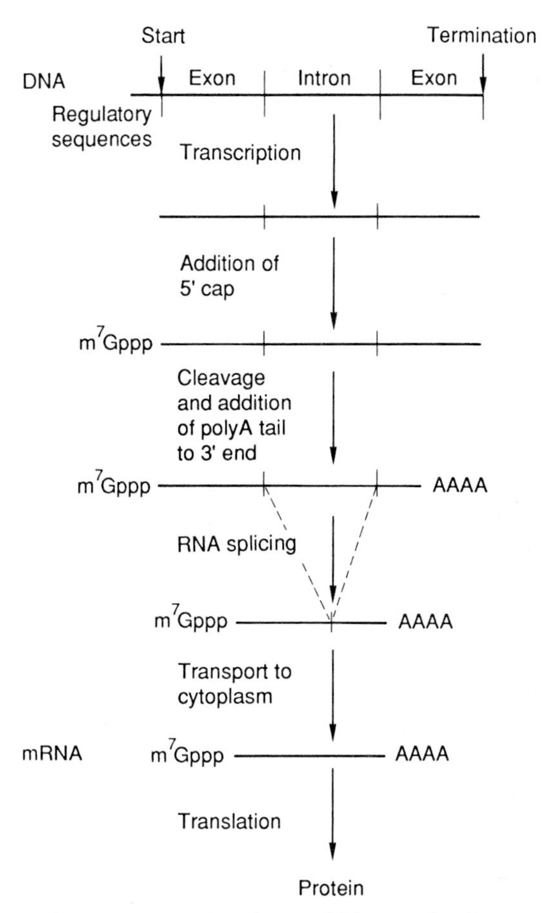

FIG 1—Stages in gene expression that could be regulated

them are actually used. Thus, for example, the production of many new proteins in the egg immediately after fertilisation and the start of embryonic development depends on the translation into protein of fully spliced mRNAs that pre-existed in the cytoplasm of the unfertilised egg but whose translation was blocked before fertilisation. This form of gene regulation is known as translational control. Similarly, splicing the protein coding regions (exons) of a single primary transcript in different combinations can produce two or more different mRNAs encoding different proteins in different tissues. This process of alternative splicing[3] is well illustrated in the single gene that encodes both the calcium

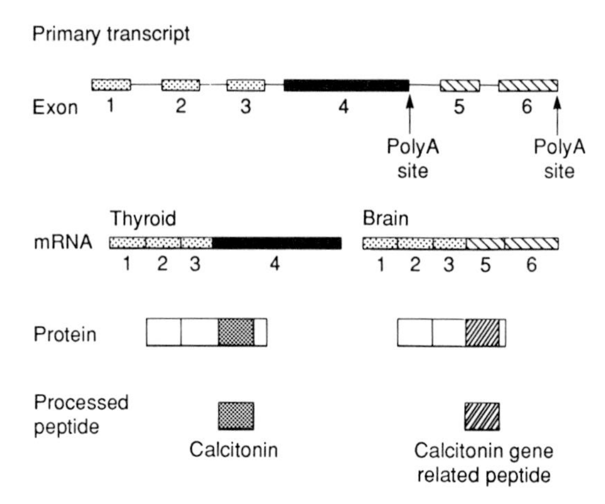

FIG 2—Alternative splicing of primary transcript of calcitonin/calcitonin gene related peptide (CGRP) gene in brain and thyroid cells. Splicing followed by proteolytic cleavage of precursor protein produced in each tissue yields calcitonin in thyroid and calcitonin gene related peptide in brain

modulating hormone, calcitonin, and the potent vasodilator, calcitonin gene related peptide (CGRP) (fig 2). Thus this gene is transcribed into a primary RNA transcript both in the thyroid gland and in the brain, but a different combination of exons are spliced together in each cell type to produce calcitonin mRNA in the thyroid and calcitonin gene related peptide mRNA in the brain.

Although there are several cases of gene regulation after the first stage of transcription, a wide variety of evidence indicates that in most cases gene regulation is achieved at the initial stage of transcription, by deciding which genes should be transcribed into the primary RNA transcript.[4] In these cases, once transcription has occurred, all the other stages in gene expression shown in figure 1 follow and the corresponding protein is produced. Thus the myosin gene is transcribed only in muscle cells, resulting in myosin being produced only in this cell type; the immunoglobulin gene is transcribed only in B lymphocytes, which produce immunoglobulin; and so on. Indeed, even in the case of calcitonin and calcitonin gene related peptide alternative splicing is acting as a supplement to transcriptional control since the calcitonin/calcitonin gene related peptide gene is transcribed only in the thyroid gland and the brain and not in other tissues. Thus the regulation of gene transcription has a critical role in the regulation of gene expression.

Regulation of transcription

For a gene to be transcribed it is necessary for specific protein factors known as transcription factors[5] to bind to particular DNA binding sites in the regulatory regions of the gene and induce its transcription by the enzyme RNA polymerase. Some of these factors are present in all cell types though others are active only in specific cells or after exposure to a particular stimulus. The combination of particular binding sites in a particular gene determines the transcription factors that bind to it, and in turn the presence or absence of these factors determines in which cell type(s) the gene is transcribed. Thus, for example, the immunoglobulin genes contain a binding site for the octamer binding transcription factor, Oct-2, in their regulatory region or promoter, upstream of the start site for transcription. The Oct-2 factor is synthesised only in B lymphocytes and hence binds only to the immunoglobulin gene promoter in B cells, resulting in the transcription of the immunoblogulin genes only in antibody producing B cells. Similarly, genes expressed only in muscle cells, such as the creatine kinase gene, contain binding sites for the MyoD transcription factor, which is present only in muscle cells. This case is even more dramatic, however, because the artificial expression of MyoD in non-muscle cells such as fibroblasts is sufficient to convert them into muscle cells, indicating that MyoD activates transcription of all the genes whose protein products are necessary to produce a differentiated muscle cell.[6]

Unlike the case of MyoD, the expression of Oct-2 alone is not sufficient to produce differentiated B cells. This is because other transcription factors that are specifically active in B cells are also involved in producing the expression of genes specific to B cells such as those encoding the immunoglobulins. One such factor is NFκB, which binds to a DNA sequence in the regulatory region of the immunoglobulin κ light chain gene. Interestingly, unlike Oct-2, the NFκB protein is present in all cell types. In most cells, however, it is present in an inactive form in which it is complexed with an inhibitory protein, resulting in it being restricted to the cell cytoplasm. In mature B cells, however, NFκB is released from the inhibitory protein and moves to the nucleus, where it can bind to

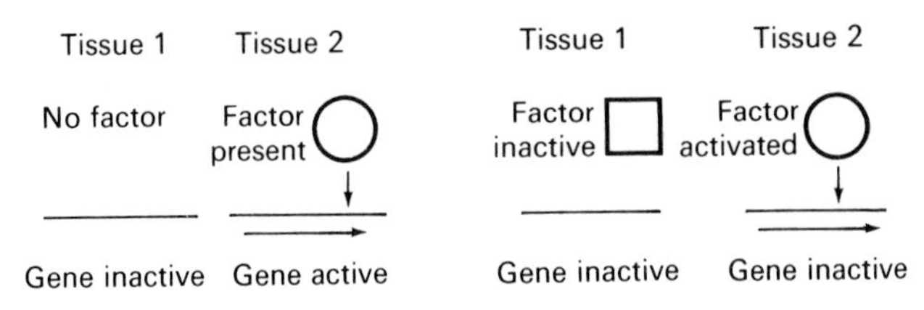

FIG 3—Transcription factors can activate gene expression in a particular tissue if (left) they are synthesised only in that tissue or (right) are present in an active form in that tissue

its DNA target sequence and activate the transcription of the immunoglobulin κ light chain gene.[7] Interestingly, this activation of NFκB also occurs when resting T lymphocytes are activated by antigenic stimulation and is the main reason for the improved growth of HIV in activated compared with resting T cells as NFκB can bind to two sites within the HIV promoter and activate viral transcription.

Hence the action of transcription factors on gene expression can be controlled not only by regulating their synthesis but also by regulating their activity (fig 3). The combination of these two processes allows transcription factors to regulate the expression of numerous different genes in different cell types.

Malregulation of gene expression in disease

Failure of transcription factor function

In view of the complex nature of gene regulation it is not surprising that it can go wrong and that a number of human diseases have now been shown to be due to defects in gene regulation. Thus one type of congenital severe combined immunodeficiency is caused by a failure of HLA class II gene transcription, resulting in the absence of these proteins. In turn this failure of transcription is dependent on the lack of a specific transcription factor necessary for the transcription of these genes.[8] Conversely in haemophilia B the particular transcription factor necessary for transcription of the factor IX gene is present, but it fails to bind to the gene promoter owing to a mutation in the DNA sequence to which it would normally bind, hence resulting in a failure of gene transcription.[9] Such defects in gene regulation can also affect steps of gene expression other than transcription. Thus the failure to produce one of the two alternatively spliced mRNAs derived from the porphobilinogen deaminase gene is the cause of one form of acute intermittent porphyria.[10]

Proto-oncogenes

As well as cases such as these, where a disease is caused by the failure to express a particular gene, malregulation of gene expression can also result in disease if it causes genes to be expressed at the wrong time or in the wrong place. This form of malregulated gene expression is central to the development of certain cancers. Thus it is now clear that most, if not all, human cancers are caused by the mutation or over-expression of certain specific cellular genes known as proto-oncogenes, which results in their conversion into cellular oncogenes capable of causing cancer.[11] Although proto-oncogenes encode many different types of cellular proteins involved in growth regulation (such as growth factors or their receptors), several (such as erbA, fos, jun, myb, and myc) encode cellular transcription factors that are involved in regulating the expression of specific genes. After the conversion of these proto-oncogenes into oncogenes by mutation or overexpression, corresponding alterations occur in the expression of the genes that they regulate, which results in cancer.

One example of this is provided by the related Fos and Jun proteins,[12] both of which encode cellular transcription factors. After treatment of cells with growth stimulating factors, Fos and Jun are synthesised and activate the transcription of specific genes whose protein products are necessary for cellular growth. Normally, however, Fos and Jun are synthesised only transiently in response to exposure to the growth factor, resulting in only a transient activation of gene expression and thereby producing the controlled proliferation, regulated by growth factor, that is characteristic of normal cells. If for any reason, however, Fos and Jun are synthesised continually, either owing to a mutation, resulting in their continual overexpression, or to infection with a virus expressing one or other of them, the cell is stimulated to grow continually even in the absence of growth factors. Such continuous uncontrolled growth is characteristic of the cancer cell.

Hence the fos and jun genes are proto-oncogenes, the products of which have a critical role in the growth of normal cells but which can be converted into oncogenes capable of transforming cells.

Moreover, in contrast to the other diseases I have discussed, in this case malregulation of gene expression and disease is caused not by failure of transcription factor function but rather by failure to regulate correctly the activity of the factor so that it is active in the wrong place at the wrong time. This indicates that, as with other cellular processes, gene expression is subject to complex regulatory mechanisms, the failure of which can be as disastrous as the failure of the basic process itself.

Conclusion

This chapter gives a brief overview of the major aspects of gene regulation mechanisms (for further details see elsewhere).[4] But much remains to be understood. For example, it remains unclear how the expression of specific genes is regulated both spatially and temporally during development so that each cell type arises in the correct place and at the correct time. It is already clear, however, that the correct regulation of gene expression is central to health and correct development and that its malregulation is involved in a number of diseases.

1 Nevins JR. The pathway of eukaryotic mRNA transcription. *Ann Rev Biochem* 1983;**52**:441–6.
2 Sharp PA. Splicing of messenger RNA precursors. *Science* 1987;**235**:766–71.
3 Latchman DS. Cell-type specific splicing factors and the regulation of alternative RNA splicing. *The New Biologist* 1990;**2**:297–303.
4 Latchman DS. Gene regulation: a eukaryotic perspective. London: Unwin Hyman, 1990.
5 Latchman DS. Eukaryotic transcription factors. *Biochem J* 1990;**270**:281–9.
6 Olson EN. Myo D family: a paradigm for development? *Genes and Development* 1990;**4**:1454–61.
7 Lenardo MJ, Baltimore D. NF-kappa B: a pleiotropic mediator of inducible and tissue-specific gene control. *Cell* 1989;**58**:227–9.
8 Reith W, Satola S, Sanchey CH, Amaldi I, Lisowska-Grospiere B, Griscelli C., *et al.* Congenital immunodeficiency with a regulatory defect in MHC class II gene expression lacks a specific HLA-DR promoter binding protein RF-X. *Cell* 1988;**53**:897–906.
9 Crossley M, Brownlee GG. Disruption of a C/EBP binding site in the factor IX promoter is associated with haemophilia B. *Nature* 1990;**345**:444–6.
10 Grandchamp B, Picat C, Mignotte V, Wilson JHP, Te Velde K, Sandkuyl L., *et al.* Tissue-specific splicing mutation in acute intermittent porphyria. *Proc Natl Acad Sci USA* 1989;**86**:661–4.
11 Bishop JM. The molecular genetics of cancer. *Science* 1987;**235**:305–11.
12 Curan T, Franza BR. Fos and Jun: the AP-I connection. *Cell* 1988;**55**:315–97.

Molecular genetics of common diseases

James Scott

Introduction

Common diseases like coronary heart disease, essential hypertension, diabetes mellitus, psychotic illness, senile dementia, cancer, and susceptibility to infectious disease tend to cluster in families. The familial aggregation of these disorders is rarely caused by a single gene defect; rather it results from the cumulative interaction of several genes with environmental factors. These disorders are therefore said to show multifactorial, or polygenic, inheritance. The risk of polygenic disease in first degree relatives is generally less than the one in four risk for mendelian recessive disorders, being about 5–15% (table I),[1] though the risk varies from one disease to another and from one family to another. Within a family the risk will depend on the severity of the disorder in the proband, the number of family members affected, and the contribution from environmental factors.

Genetic epidemiology and genetic modelling

For any disease we need to establish whether its cause has a genetic component. Family, twin, and adoption studies help to establish whether there is a genetic component, the size of that component, and the risk of disease in the relatives of a proband. Knowing the risk of a disease in relatives and curve fitting variables, such as blood pressure or plasma cholesterol concentrations, for affected compared with normal families (commingling analysis) may suggest a mode of inheritance. For example, a normally distributed curve is usually indicative of polygenic inheritance, whereas skewedness or bimodality may suggest the

TABLE I—*Risks for common polygenic diseases of adults*[1]

Disorder in proband	Risk (%) for first degree relatives
Coronary heart disease	8 for male relatives 3 for female relatives
Hypertension	10
Diabetes mellitus	5–10
Epilepsy	5–10
Manic depressive psychosis	10–15
Schizophrenia	15
Psoriasis	10–15
Thyroid disease	10

(after Goldstein and Brown[1])

influence of major genes or of major environmental factors. Analysing the segregation of a disease within families (complex segregation analysis) with computer programs like pedigree analysis package (PAP) also helps to establish the mode of inheritance of a trait—that is, whether it is likely to be caused by a major gene showing either dominant or recessive inheritance or is caused by several genes (that is, whether inheritance is oligogenic or polygenic), or whether environmental factors are largely responsible.

Number of genes implicated in polygenic disease

After establishing whether a disease has a genetic component and, if so, the size of this component, it is necessary to ascertain what genes potentially participate in a disease. For a complex disorder such as coronary heart disease, in which plasma lipoproteins, the coagulation system, and the cellular elements of the blood and the arterial wall play a part, the number of genes may be large. One component of this problem illustrates its complexity. Genes that affect plasma lipids have been the subject of much research, so much is known about some of them, including those coding for: apolipoproteins that carry cholesterol and triglyceride in the circulation, lipid transfer proteins, receptors for apolipoproteins, or key enzymes affecting lipid metabolism (table II). This list includes several genes that, in mutant form, seriously perturb blood lipid concentrations. It is by no means comprehensive. At least 200 genes have been estimated to play a part in controlling the uptake of cholesterol by the gut, its metabolism in the plasma, liver, and peripheral cells, and its excretion from the body. Many of these genes have not been identified. The issues here are, firstly, to establish which of the known genes affect plasma cholesterol concentrations. The extent to which many known genes contribute, if at all, has still to be established. Secondly, new loci that contribute to lipid metabolism must be identified. Thirdly, genes that contribute to other aspects of the problems of atherogenesis must be found. The same questions pertain to other common diseases with a complex genetic component.

Genetic variation and susceptibility to disease

DNA sequence variation occurs about every 200 and 500 base pairs. Thus most genes can be expected to show variation in every human population. Sequence variants (mutations) that occur in more than 1% of the population are generally called polymorphisms and those occurring in less than 1% are termed rare alleles (variation at a single gene locus). For most individuals the risk of developing a disease will depend on a complex interaction between common alleles, which each have small effects that combine additively. Superimposed on the population distribution contributed by polymorphic alleles will be the influence of rare alleles.

POLYMORPHISMS THAT PREDISPOSE TO POLYGENIC DISEASE

Common alleles, or polymorphisms, form the basis of human diversity, including our ability to handle environmental challenges, such as exposure to infectious agents or chemical carcinogens or excessive consumption of saturated fat and cholesterol. Such alleles are likely to have become more prevalent since the original mutation occurred, because of selection acting on variants that confer advantage in the heterozygous state. For example, the genetic polymorphisms that contribute to diseases like coronary heart disease, essential hypertension, and diabetes mellitus almost certainly have a high prevalence in the population. Perhaps these variants, which once conferred advantage by maintaining blood pressure and concentrations of glucose and cholesterol at times when food was scarce for our hunter-gatherer ancestors (the thrifty genotype hypothesis of JV Neel), now respond to overnutrition by predisposing to the common diseases that affect people today. Is it surprising, in evolutionary terms, that modern humans so crave sugar, salt, and fat?

Polymorphisms, population association, and linkage disequilibrium

How are we to establish whether a particular gene contributes to the genetic variation in a trait such as blood pressure or plasma cholesterol concentration? Genetic epidemiology has provided statistical methods for measuring the effect of genetic variation on a phenotypic trait in a population (the so called measured genotype approach). For example, functional polymorphism of apolipoprotein E (encoded on chromosome 19q13) can be assessed by isoelectric focusing or by the polymerase chain reaction (PCR) and the use of allele specific oligonucleotides. In the normal population these three polymorphisms have a substantial effect on the normal variation in plasma lipid concentrations (table III); they account for 16% of genetic variation in cholesterol concentrations in the population.[2] These small changes in plasma cholesterol concentration are sufficient to produce a substantial enrichment of the E2 allele with advancing age and a decrease of the E4 allele in the aging population, presumably as a result of an increased number of deaths from myocardial infarction. Yet the effect of these polymorphisms on individual lipid concentrations and risk of coronary heart disease is small.

The exception to this is when homozygosity of the E2 allele occurs in association with diabetes mellitus, hypothyroidism, or with another genetic defect of lipid metabolism. This leads to the condition of dysbetalipoproteinaemia (type III hyperlipidaemia), which is characterised by profound hypertriglyceridaemia, moderate hypercholesterolaemia, and premature coronary heart disease. This condition is caused by defective clearance of chylomicron

TABLE II—*Key proteins associated with lipid metabolism*

Class	Protein	Function
Plasma apolipoproteins	Apolipoprotein A I	HDL structural protein, LCAT activation
	Apolipoprotein A II	HDL structural protein
	Apolipoprotein A IV	Unknown
	Apolipoprotein B 100	VLDL assembly and secretion; ligand for LDL receptor
	Apolipoprotein B 48	Chylomicron assembly and secretion
	Apolipoprotein C I	Unknown
	Apolipoprotein C II	Lipoprotein lipase activation
	Apolipoprotein C III	Lipoprotein lipase inactivation
	Apolipoprotein E	Ligand for chylomicron remnant and LDL receptors

TABLE II—*Key proteins associated with lipid metabolism* (continued)

Class	Protein	Function
Enzymes	AMP-activated protein kinase	Inhibits HMG CoA reductase, acetyl-CoA carboxylase, and hormone sensitive lipase
	Cholesterol 7-α hydrolase	Cholesterol conversion to bile acids
	Cholesterol-ester hydrolase	Intracellular cholesteryl-ester hydrolysis
	Endothelial lipoprotein lipase	Lipolysis of triglyceride-rich chylomicrons and VLDL
	Fatty acid synthetase	Fatty synthesis
	Fatty ACAT	Cellular cholesterol esterification
	Hepatic triglyceride lipase	Lipolysis of remnants and HDL
	Hormone sensitive lipase	Hydrolysis of intracellular triglyceride in fat and muscle cells
	HMG CoA reductase	Rate limiting enzyme of cholesterol synthesis
	LCAT	Plasma cholesterol esterification
	Phosphatidic acid phosphohydrolase	Phospholipid synthesis
	5-, 12- and 15-lipoxygenase	Arachidonic acid oxygenation and oxidisation of LDL
Lipid transfer proteins	Acetyl-CoA carboxylase	Fatty acid synthesis
	CETP	Cholesterol transfer from HDL to VLDL and LDL
	Hepatic fatty acid binding protein	Intracellular fatty acid transport
	Intestinal fatty acid binding protein	Intracellular fatty acid transport
Receptors	MTP	Lipid loading of nascent lipoproteins
	Apolipoprotein E	Remnant clearance
	LDL	LDL and remnant clearance
	VLDL receptor	VLDL clearance
	Scavenger receptor	Oxidised LDL clearance by macrophage

HDL = high density lipoprotein, LCAT = lecithin-cholesterol acyl-transferase, LDL = low density lipoprotein, VLDL = very low density lipoprotein, HMG = hydroxymethyl glutaryl, CoA = coenzyme A, ACAT = acyl-CoA cholesterol acyl-transferase, CETP = cholesteryl-ester transfer protein, MTP = microsomal triglyceride transfer protein.

TABLE III—*Effect of apolipoprotein E alleles on cholesterol concentrations*

Allele	Position	Mutation	Receptor binding (%)	Prevalence (%)	Cholesterol mg/dl	Cholesterol mmol/l
E4	112	Arginine	100	15	+0·32	(+13)
E3	112	Arginine→Cysteine	100	72	0	(0)
E2	158	Arginine→Cysteine	2	13	−0·21	(−8)

Data from Davignon and Mahley (see Scott[2])

remnants (caused by the E2 allele, which does not bind to the low density lipoprotein or chylomicron remnant receptor) combined with another condition that produces hyperlipidaemia.

A remarkable size polymorphism is shown by the apolipoprotein(a) gene (apo(a)).[3] Apolipoprotein(a) resembles the precursor of the enzyme, plasmin, which is responsible for the hydrolysis of fibrin clots. Apolipoprotein(a) is secreted tightly associated with apolipoprotein B 100 from the liver, where the two molecules together form a lipoprotein particle designated Lp(a) lipoprotein. The apo(a) gene has a variable number of repeats of the region coding for kringle (a disulphide bonded domain shaped like a Danish cake) IV. This gives rise to extreme size variation from 280 to 830 kDa. The size is inversely correlated with Lp(a) lipoprotein levels in the circulation, and individuals with the small protein have much higher concentrations and are at much greater risk of coronary heart disease. The apo(a) gene size polymorphism accounts for more than 90% of the variation of Lp(a) lipoprotein levels in the population. Though the cause of the increased likelihood of disease is not certain, Lp(a) lipoprotein probably in some way interferes with normal endothelial cell fibrinolysis and promotes a procoagulant state. Raised blood Lp(a) lipoprotein concentrations with raised cholesterol concentrations synergise in increasing the risk of coronary heart disease.

Electrophoresis certainly underestimates protein polymorphism: most proteins do not show size variation, isolectric focusing does not detect uncharged amino acid changes, and only four of the 20 essential amino acids are charged. Moreover, variation in the DNA sequences concerned with the regulation of gene expression will not be detected. In the absence of functional protein polymorphisms, population association studies have been performed using biallelic restriction fragment length polymorphisms (RFLPs) (highly polymorphic repeats are not useful because their stability in time is uncertain and they are hard to read on gels). Most usually, restriction fragment length polymorphisms do not mark functional change in the DNA, but may be closely linked to genetic variation causing a change in phenotype. Association studies with restriction fragment length polymorphisms usually depend on linkage disequilibrium. Linkage disequilibrium is said to exist when two markers, or a marker and a trait, are found in association in a population at a frequency greater than would be expected by chance alone (that is, random association). Usually this implies that the two mutations are closely linked on the same chromosome and cosegregate in family studies. The association of disease with a restriction fragment length polymorphism implies that natural selection has acted in the heterozygous state to increase the frequency of an allele carrying the marker restriction fragment length polymorphism and the casual mutation. For this to have occurred appreciable inbreeding must have taken place in the founder population, in which causal mutation occurred, during a period of population stability after the mutant allele and marker polymorphism first appeared together. Amino acid charge variants are more likely to be direct in their effect, but may also of course mark the mutation causing the trait through linkage disequilibrium.

The value of population association studies that depend on linkage disequilibrium are well illustrated by alleles in the HLA complex.[4] The HLA genes are located on chromosome 6p21 and are found at four closely linked loci (A, B, C, and D). The products of these genes are concerned with the presentation and recognition of foreign antigens. Multiple alleles (20 or more, of which each individual may inherit any two) exist for each of the four genes. Certain HLA genes occur together more often than would be predicted by chance alone. A classic example is the finding that in north European white populations HLA A1 and B8 occur together at least four times as often as would be expected by chance because they are in linkage disequilibrium. This association was probably selected for during the evolutionary history of the north European white population; it may have conferred protection against a disorder such as plague. An example of such selection may be seen among the ancestors of Dutch settlers in Surinam: most of the original settlers succumbed to a typhoid epidemic, and only those with particular HLA genes survived.

Another important HLA association is that of an HLA class I and a class II haplotype, which is common in west Africa but rare in other regions and which confers protection against severe malaria.[5] This and a plethora of other HLA associations with immunologically mediated and infectious disease strongly support the view that the extraordinary polymorphism of the major histocompatibility complex (MHC) genes has evolved through natural selection to protect organisms against infectious agents. As the molecular basis of antigen presentation is defined better, the role of structural variation in producing these associations is being understood increasingly.

There are many other examples of population association caused by linkage disequilibrium. In addition to helping investigation of the apo E gene, population association has been useful in showing that genetic variation at other key loci associated with lipid metabolism has an appreciable impact on the genetic variation of blood lipid concentrations (table II). Similarly, genetic variations of the fibrinogen gene and of the clotting factor VII and VIII genes contribute to the concentrations of these proteins in the blood, and genetic variation at the insulin locus influences the risk of developing insulin dependent diabetes mellitus.[4,6–8] Increasingly, population association studies show the principle that genetic variation of the major loci associated with a disease will influence the expression of the phenotype.

With the availability of better strategies for screening for mutation, the genetic changes responsible for population association between loci and disease are susceptible to analysis. Population association is a prelude to, rather than substitute for, identifying the precise mutations in the DNA that predispose to genetic variation. Techniques such as chemical cleavage of DNA and single stranded conformational analysis of polymorphisms in the DNA have made screening for mutations relatively straightforward. Once a mutation has been identified, polymerase chain reaction strategies and oligonucleotide probes specific for alleles make the study of its impact on a disease process a straightforward matter.

RARE ALLELES

Rare alleles also contribute to genetic diversity and susceptibility to disease. Indeed, they are the substrate of mendelian genetic disease. Though the effect of such changes may be devastating to individuals and cause genetic disease, in population terms a single rare mutation is insignificant. However, the cumulative effect of each of the rare mutations occurring at each of the loci that confer the risk of a disease may be substantial. Thus, defects of the low density lipoprotein (LDL) receptor and apolipoprotein B (apo B) genes both cause serious hypercholesterolaemia, and though the prevalence of each disorder in itself is only about 1/500, each affects 1/50 people with low density lipoprotein cholesterol concentrations above the 90th percentile.[2 9 10] Many different mutations of the low density lipoprotein receptor gene (chromosome 19p13) have been shown to cause familial hypercholesterolaemia, whereas a particular single defect (apolipoprotein B arginine$_{3500}$ →glutamine) of the apo B gene (chromosome 2p24-p23) has also been associated with a similar codominantly inherited form of hypercholesterolaemia.

Homozygosity for defects of the enzyme cystathionine β-synthase (chromosome 21q22) causes the rare disorder, homocysteinuria.[11] People homozygous for homocysteinuria have ocular, skeletal, and neurological problems and a high risk for premature atherosclerosis and venous or thromboembolism. Heterozygosity for this disorder, which occurs in 1–2% of the population, is an increased risk for premature atherosclerosis. High levels of homocysteine in the blood are toxic to vascular endothelium, may potentiate oxidative damage to low density lipoprotein, and promote thrombosis. These rare alleles contribute appreciably to the burden of atherosclerosis.

Similarly, the dementia called spongiform encephalopathy—once thought to be a slow viral disease of sheep (scrapie) and associated with canibalism in New Guinea (kuru)—has now been found to be heritable (familial Creutzfeldt-Jakob disease, Gerstmann-Straussler-Scheinker syndrome, and fatal familial insomnia) and has now been shown to be strongly correlated with rare mutations of a gene (the prion gene) that codes for a cell surface glycoprotein. Remarkably, mutations of this gene can induce a transmissible, conformational change within the prion protein that leads, in the absence of infective nucleic acid, to the disorder.[12 13] Now homozygosity for polymorphism of the prion gene has been found to produce Creutzfeld-Jakob disease.[12]

Linkage studies with candidate genes

How can we establish whether a rare allele of a specific gene causes a particular disorder in a family or group of families? The term "candidate" is used when, for functional reasons, a particular gene has a strong possibility, if defective, of causing a disease. As an example, let us again consider high blood cholesterol concentration. This could be due to a defect of either the low density lipoprotein receptor or of its ligand, apolipoprotein B. If genetic modelling suggests that high blood cholesterol concentration with tendon xanthomas and high risk of myocardial infarction (familial hypercholesterolaemia) is likely to be caused by the dominant effect of a major gene then it would be reasonable to test whether a mutation of the low density lipoprotein receptor gene causes this disorder. This can be tested with probes for the low density lipoprotein gene, or nearby flanking markers, and a computer program for linkage analysis. If a mutation of this gene causes familial hypercholesterolaemia, then cosegregation of a specific allele of the gene (marked by a restriction fragment length polymorphism or other polymorphic repeat) with the disease will occur within a large affected family. If cosegregation does not exist with the allele then the low density lipoprotein receptor gene can, by and large, be eliminated as a cause of the disorder in this family. On the other hand, if consistent cosegregation does exist a causal role for the gene in the disease is strongly supported.

In fact, the low density lipoprotein was identified by classic biochemical methods because the uptake of low density lipoprotein by cultured fibroblasts was shown to be abnormal in families with severe hypercholesterolaemia. If a single very large family was studied, in which the familial form of hypercholesterolaemia was caused by a defect of either the low density lipoprotein receptor gene or the apo B gene, then linkage would be established. However, if data on several families with severe hypercholesterolaemia were pooled and some showed defects of the low density lipoprotein receptor gene, some of the apo B gene, and some of as yet unknown genes that cause severe hypercholesterolaemia then linkage would not be established. The same phenotypic abnormality being caused by defects at several distinct genes, which is described as locus heterogeneity, is a serious problem in linkage studies.

Linkage studies and positional cloning

How can the new loci that contribute to disease be identified, when the phenotype provides no information about the locus or the biochemical abnormality responsible for the disease? If such a disorder cannot be linked to a candidate gene then the entire genome may have to be screened for linkage. This strategy has been much simplified by the identification of variable nucleotide tandem repeats (VNTRs) and simple dinucleotide, trinucleotide, and tetranucleotide repeats. With such highly polymorphic markers chosen to be spaced at a genetic distance of 10–20 centimorgans, it should be possible to ascertain whether there is linkage to a locus that harbours the gene responsible for a disorder and to use the power of modern molecular genetics to clone the genes.

This approach has been successful in an increasingly growing list of monogeneic disorders. It is exemplified by successfully identifying the locus and gene for cystic fibrosis, and this has been followed by characterising many of the mutants causing the disease and by cellular and molecular studies aimed at defining the function of the plasma membrane transporter protein that the gene encodes. The identification of the loci responsible for adult polycystic kidney disease (chromosome 16p13)[14] and the Romano-Ward syndrome (chromosome 11p15)[15] of QT interval prolongation on electrocardiogram, which predisposes to sudden death from ventricular tachycardia, and the disclosure that mutations of the β myosin heavy chain gene cause hypertrophic obstructive cardiomyopathy[16] are good examples of studies of relatively rare disorders which can increase our knowledge of more common defects.

In polygenic disease a common phenotype is produced by the interaction of several genes so the use of this approach is much more difficult. If, however, clinical material is chosen carefully to maximise single gene effects, there is no reason why this approach should not work. The study of late and early onset Alzheimer's disease is a clear example of the success of this strategy.[17] Another example of this approach is the identification of the apolipoprotein AI/CIII/AIV gene cluster as predisposing to familial combined hyperlipidaemia in some families.[18] For many diseases this may be the only approach, but if too much heterogeneity exists then this strategy is unlikely to succeed.

Classic genetics

The strategy of positional cloning may be simplified by clues from classic genetic studies, in which chromosome deletions or translocations may provide clues that identify the occult loci responsible for genetic disease. Thus genetic abnormalities such as trisomy 21 have pointed to genes playing a part in the molecular basis of Alzheimer's disease, as Down's syndrome patients surviving into their fifth and sixth decades almost invariably develop brain lesions typical of this disorder.[16] Interestingly, the gene that codes for Alzheimer's disease, β amyloid protein, has been located to chromosome 21q21. The gene codes for a transmembrane glycoprotein that may, under certain circumstances, undergo abnormal proteolytic cleavage and interfere with homeostatic mechanisms that protect against abnormal protease activation in the nervous system. Mutations of this gene, which lead to dominant inheritance of Alzheimer's disease, have now been identified. In sporadic Alzheimer's disease, for which no genetic basis has yet been found (this is difficult in a disease of such late onset), common polymorphisms of the amyloid precursor protein may possibly be found to predispose to the disease. Thus classic genetics, linkage studies, molecular genetics, and biochemistry have come together to identify a main component associated with Alzheimer's disease. Greater understanding of the pathogenesis of such disorders will probably lead to uncovering other genetic components that predispose to the prevalent problem of dementia (10% of people over 65 and 50% of those over 85).

Congenital heart disease is also common in patients with Down's syndrome; this may provide a clue to localisation of one of the genes that lead to this common birth defect. Classic genetics has also done much towards an understanding of the molecular basis of certain forms of cancer. For example, a chromosome 5q translocation was instrumental in pinpointing the gene in a kindred with familial polyposis coli and for enhancing our understanding of the defects leading to less obviously hereditary forms of colonic cancer.[19] Classic genetics still has much to offer molecular genetics.

Problems with linkage studies

In studying the genetics of common diseases, linkage studies have run into problems. This has been for two principal reasons. Firstly, phenotype definition may be difficult. Defining the phenotype is particularly difficult in psychiatric disorders, for which there are no biochemical or medical markers and, until recently, no definitions of disease had been generally agreed. Problems of definition also exist in considering variables such as blood lipid or glucose concentrations or blood pressure. These variables are measurable traits, which are distributed in the population according to genetic and environmental factors. Thus though the extremes of the distribution of blood cholesterol concentrations may be caused by major genes, it is difficult to separate important gene effects from the multiple genetic and environmental factors that determine the distribution of cholesterol concentrations in the population. Secondly, locus heterogeneity as described above may confound the ascertainment of linkage.

Nowhere has the problem of linkage studies been more apparent than in studying the major psychiatric disorders.[20] Manic depressive illness was reported to be linked to the Harvey ras oncogene and insulin loci on chromosome 11p15 in an old order Amish kindred. These genes are close to the one encoding tyrosine hydroxylase—an important candidate gene because of its influence on dopamine metabolism. This linkage has now been firmly refuted. Problems in the original study were caused partly by incorrect definition of the phenotype, partly by reduced penetrance of the disorder, and partly by misreading of the highly polymorphic markers used for the linkage study. Manic depressive illness has also been linked by several markers to chromosome Xq28 near the fragile site in some families.

Linkage studies of schizophrenia have been even more fraught than those of affective illness. Initially, classic genetic studies suggested a candidate locus for schizophrenia on chromosome 5. This locus appeared to be confirmed in linkage studies identifying the region of chromosome 5q11-q13 in Icelandic families, but the linkage has not been confirmed in British, American, or other Scandinavian families.

Thus, so far in the main psychiatric disorders, linkage assignments are controversial and at best tentative. Although major genetic components exist for affective disorders and schizophrenia, the mode of their inheritance is not yet clear.

OLIGOGENIC OR POLYGENIC DISORDERS

Affected sib and relative pair methods

If genetic epidemiology suggests that the mode of inheritance of a particular disorder is not clear or is due to the interaction of several genes, then it is inappropriate to perform classic linkage studies in large families or in pooled small families. In this case the affected sib pair or relative pair method has the advantage that linkage can be detected if a disease shows non-mendelian inheritance because of several interacting genes, provided that the origin of each of the four parental chromosomes can be identified in the affected relative pairs—so called identity by descent. This approach is less useful in dominant disorders or where there is genotype heterogeneity and has so far only been applied successfully to candidate genes such as the HLA locus in insulin dependent diabetes because the power of the approach decreases rapidly with increased genetic distance away from the affected locus.

Animal models

The study of inbred strains of mice and of other animal models of genetic disease has provided an invaluable tool for analysing the complex patterns of inheritance found in polygenic disease and for identifying occult loci. Most recently, the genetics of insulin dependent diabetes and of essential hypertension have been the subject of intensive study using mouse and rat models, respectively.

Non-obese diabetic (NOD) mice that spontaneously develop diabetes mellitus have remarkable similarities to humans with insulin dependent diabetes. In both species there is autoimmune islet cell destruction, autoantibodies raised against B cell components, and defects in T cell activity, as well as susceptibility genes in the major histocompatibility complex. Non-obese diabetic mice share the same HLA association, with serine at position 57 in the β chain.[21 22] In addition to the mouse major histocompatibility complex locus on mouse chromosome 17, linkage studies in the non-obese mice have identified loci designated idd-3, idd-4, and idd-5 on chromosomes 3, 11, and 5. These probably correspond to loci on human chromosome 1 or 4 for idd-3 and 17 for idd-4. Mapping shows idd-5 close to the interleukin 1 receptor, which mediates resistance to bacterial and parasitic infection and affects the function of macrophages.[23] Susceptibility to infection is considered to be a main component in the aetiology of islet destruction.

There is also a mouse model of non-insulin dependent diabetes.[24] On mouse chromosome 4 an autosomal recessive mutation is associated with profound obesity and hyperphagia, increased metabolic efficiency, and insulin resistance—interbreeding with different mouse strains modulates this phenotype greatly.

Chromosome mapping of genes affecting hypertension have focused on the spontaneously hypertensive rat as a stroke-prone model of human disease.[25 26] In this animal an important locus has been identified on rat chromosome 10. This region is closely linked to the rat gene encoding angiotensin converting enzyme (ACE). This enzyme plays a major part in blood pressure homeostasis and is an important target of antihypertensive drugs. In humans this gene resides on chromosome 17q23. Other loci associated with hypertension have been found on rat chromosomes 18 and X. Other rat strains with raised blood pressure should prove valuable in mapping further disease loci.

The future

Before the year 2000 molecular genetics will have been established, physical maps of the entire human genome made, and much of the genome sequenced.[4] Most of the genes implicated in common polygenic diseases of adults and common birth defects of children will have been characterised and the mutant alleles that predispose to disease will have been identified fully. By using procedures for amplifying DNA and using allele specific oligonucleotide probes, screening for polymorphisms associated with a wide variety of common diseases will be possible in a single reaction. The disease risk profiles of individuals will be available before they are born.

Screening in early life or antenatally for genes that predispose to common ailments in later life has clear advantages, but poses ethical and practical problems. For a disorder that can be prevented, such as coronary heart disease, it is a remarkable bonus to know and to treat early. Predictive screening may lead to the modification of life style or the introduction of specific treatment. Difficulties can be envisaged, however: decisions about termination of pregnancy, lifestyle, health insurance, or occupation could be based on relatively small and ill judged genetic risks. And are we willing to pay? For monogenic disorders it is certainly cheaper to screen antenatally than to care for chronically handicapped people. For common diseases of adults, the answer is also likely to be "yes"—for example, in terms of health care alone, coronary heart disease has been estimated to cost £1 billion a year in Great Britain, and this does not take into account the even greater cost to industry and to commerce. The bill could probably be halved by appropriate preventive measures.

1 Goldstein JL, Brown MS. Genetic aspects of disease. In: Wilson JD, Braunwald E, Isselbacher KJ, Petersdorf RG, Martin JB, Fauci AS, *et al*, eds. *Harrison's principles of internal medicine*. 12th ed, vol 1. New York: McGraw-Hill, 1991;21–32.
2 Scott J. The molecular and cell biology of apolipoprotein-B. *Mol Biol Med* 1989;**6**:65–80.
3 Scott J. Lipoprotein(a): thrombotic and atherogenic. *BMJ* 1991;**303**:663–4.

4 Scott J. Molecular genetics of common diseases. *BMJ* 1987;**295**:769–71.
5 Hill AVS, Allsopp CEM, Kwiatkowski D, Anstey NM, Twumasi P, Rowe PA, *et al.* Common West African HLA antigens are associated with protection from severe malaria. *Nature* 1991;**352**:595.
6 Bell GI, Wu S-H, Newman M, Fajans SS, Seino M, Seino S, *et al.* Diabetes mellitus: identification of susceptibility genes. In: Lindsten J, Pettersson U, eds. *Etiology of human disease at the DNA level.* New York: Raven Press, 1991:93.
7 Julier C, Hyer RN, Davies J, Merlin F, Soularue P, Briant L, *et al.* Insulin-IGF2 region on chromosome 11p encodes a gene implicated in HLA-DR4-dependent diabetes susceptibility. *Nature* 1991;**354**:155–9.
8 Fowkes FGR, Connor JM, Smith FB, Wood J, Donnan PT, Lowe GDO. Fibrinogen genotype and risk of peripheral atherosclerosis. *Lancet* 1992;**339**:693–6.
9 Breslow JL. Genetic basis of the lipoprotein disorders. *J Clin Invest* 1989;**84**:373–80.
10 Breslow JL. Lipoprotein transport gene abnormalities underlying coronary heart disease susceptibility. *Annu Rev Med* 1991;**42**:357–71.
11 Clarke R, Daly L, Robinson K, Naughten E, Cahalane, Fowler B, *et al.* Hyperhomocysteinemia: an independent risk factor for vascular disease. *N Engl J Med* 1991;**324**:1149–55.
12 Palmer MS, Dryden AJ, Hughes JT, Collinge J. Homozygous prion protein genotype predisposes to sporadic Creutzfeldt-Jakob disease. *Nature* 1991;**352**:340–2.
13 Weissman C. A "unified theory" of prion propagation. *Nature* 1991;**352**:679–83.
14 Reeders ST, Breuning MH, Davies KE, *et al.* A highly polymorphic DNA marker linked to adult polycystic kidney disease on chromosome 16. *Nature* 1985;**317**:542–4.
15 Keating M, Atkinson D, Dunn C, Timothy K, Vincent GM, Leppert M. Linkage of a cardiac arrhythmia, the long QT syndrome, and the Harvey *ras*-1 gene. *Science* 1991;**252**:704–6.
16 Tanigawa G, Jarcho JA, Kass S, Solomon SD, Vosberg H-P, Seidman JG, Seidman CE. A molecular basis for familial hypertrophic cardiomyopathy: an a/B cardiac myosin heavy chain hybrid gene. *Cell* 1990;**62**:991–8.
17 Yankner BA, Mesulam M-M. Beta-amyloid and the pathogenesis of Alzheimer's disease. *N Engl J Med* 1991;**325**:1849–57.
18 Wojciechowski AP, Farrall M, Cullen P, Wilson TME, Bayliss JD, Farren B, *et al.* Familial combined hyperlipidaemia linked to the apolipoprotein AI-CIII-AIV gene cluster on chromosome 11q23-q24. *Nature* 1991;**349**:161–4.
19 Stanbridge E. Human tumor suppressive genes. In: Campbell A, Baker BS, Jones EW, eds. *Annual review of genetics.* Palo Alto, USA: Annual Reviews Inc, 1990:615.
20 Ciaranello RD, Ciaranello AL. Genetics of major psychiatric disorders. *Annu Rev Med* 1991;**42**:151–8.
21 Todd JA. Genetic control of autoimmunity in type 1 diabetes. *Immunol Today* 1990;**11**:122–9.
22 Todd JA, Aitman TJ, Cornall RJ, Ghosh S, Hall JRS, Hearne CM, *et al.* Genetic analysis of autoimmune type 1 diabetes mellitus in mice. *Nature* 1991;**351**:542–7.
23 Cornall RJ, Prins J-B, Todd JA, Pressey A, DeLarato NH, Wicker LS, *et al.* Type 1 diabetes in mice is linked to the interleukin-1 receptor and *Lsh/Ity/Bcg* genes on chromosome 1. *Nature* 1991;**353**:262–5.
24 Bahary N, Leibel RL, Joseph L, Friedman JM. Molecular mapping of the mouse *db* mutation. *Proc Natl Acad Sci USA* 1990;**87**:8642–6.

25 Hilbert P, Lindpaintner K, Beckmann JS, Serikawa T, Soubrier F, Dubay C, *et al*. Chromosomal mapping of two genetic loci associated with blood-pressure regulation in hereditary hypertensive rats. *Nature* 1991;**353**:521–9.
26 Jacob HJ, Lindpaintner K, Lincoln SE, Kusumi K, Bunker RK, Mao Y-P, *et al*. Genetic mapping of a gene causing hypertension in the stroke-prone spontaneously hypertensive rat. *Cell* 1991;**67**:213–24.

Genes and cancer

Richard G Vile, Myra O McClure, Jonathan N Weber

It is clear that, during development from a single cell zygote to a multicellular organism, a critical balance must be maintained between cell numbers (the ability of individual cells to proliferate) and cell specialisation (their concomitant ability to differentiate). Without sufficient levels of proliferation there will be too few cells to carry out the functions specific to their lineage; without appropriate degrees of differentiation within the lineage the functional specialisation of the tissue will be impossible to maintain. The ability of a cell to proliferate is intricately coupled with, and generally inversely related to, its ability to differentiate, as each phenotype is ultimately determined by activation of separate programmes of gene expression. It is from disruption of this fine balance of proliferative and differentiative genetic programmes that tumours inevitably arise.

A cancer may be viewed as a population of cells that has progressed a certain distance in its maturation pathway, but in which the processes of proliferation and differentiation have become uncoupled, and that, critically, is no longer able to complete its programme of differentiation. This population of cells is usually derived from divisions of a single cell that has acquired an accumulation of damaging changes in its genes. Cancer cells are said to express the fully transformed phenotype when they have acquired the ability to grow continuously, free from the normal inhibition usually imposed by their nearest neighbours (contact inhibition), and when they have become malignant. A malignant tumour consists of fully transformed cells that can invade adjacent tissues and spread (metastasise) to other sites in the body to form secondary tumour growths. Most cells can be diverted from their normal differentiation programmes within most of the compartments of the body and at many different stages between stem cell and the fully differentiated state.

Cancer as a disease of genes

For many years it has been realised that damage to the DNA of a cell (mutation) is associated with the changes that lead to cancer (carcinogenesis). At the turn of the century, Boveri identified the chromosomes as the storage place of the cell's genetic material and proposed chromosome imbalance as a major contributory factor to cancer development.[1] It has been shown repeatedly that most carcinogens are also active mutagens and the ability to cause damage to DNA correlates in most cases very well with the ability to induce cancer. Up to 70% of human cancers are due to the action of chemical carcinogens, which can often be shown to be mutagenic in vitro. The most notable example of this is lung cancer, which is aetiologically linked with the action of activated polycyclic hydrocarbons. As epoxides, these form adducts with DNA, which cannot be adequately repaired and thereby introduce mutations.

However, the conversion of a normal cell to a malignant cancer never occurs in a single step and cannot be attributed solely to the mutation of a single gene. Rather, there must be a series of changes in the properties of the collection of cells that make up the developing tumour,[2,3] and the evolution of a tumour towards an ever more malignant phenotype is a common clinical experience. The behaviour and severity of any cancer is therefore decided by a multifactorial range of genetic changes. As cancer is a disease of proliferating cells, it was not surprising to find that many of these mutations affect genes that control the rate of cellular proliferation and the ability of a cell to differentiate.

The oncogenes

It was believed initially that the genetic mutations responsible for cancer caused a deletion of essential regulatory genes restraining cell growth. However, the discovery of retroviruses radically changed these ideas on the alterations that occur to genes in cancer, led to the discovery of oncogenes, and gave a new perspective of cancer as being caused by genes that actively promote uncontrolled growth. In 1911, Peyton Rous described a transmissible agent that could pass tumours between chickens. It was subsequently shown that this agent was a retrovirus that harboured a specific gene, v-src, which was itself responsible for the tumours. More retroviruses were discovered that could transform cells to rapid and uncontrolled growth when grown in culture, and these too contained genes implicated in carcinogenesis. The term oncogene was coined to denote those viral genes that might have a role in converting normal cells to cancerous ones, and the action of the viral proteins was envisaged as helping to push the cell into a proliferative (cancerous) cycle.

In 1976, sequences of DNA related to the retroviral oncogene v-src were shown to be present in the DNA of normal uninfected chicken cells.[4] The cellular homologues of other, v-onc, sequences were shown subsequently. Hence it was recognised that some of the genes linked to cancer exist within normal cells before infection by retroviruses occurs. Elucidation of the life cycle of the retroviruses showed that they can infrequently hijack—transduce—incomplete portions of cellular genes into their own genetic material, which leads to the damage of the normal genes in such a way that they no longer function correctly. The cellular gene may either be placed under virally determined transcriptional control (both quantitatively and temporally), or sustain critical mutations to the coding sequence such that the function of the protein product is altered, or both.[2] If the transduced gene plays a central part in controlling growth and differentiation, these changes in structure and expression may contribute to the transformation of an infected cell.

Most non-viral oncogenes seem to be altered forms of cellular genes that encode proteins that participate in the pathways of cellular proliferation. The normal, intact cellular genes are known as proto-oncogenes. Cells receive signals to proliferate via growth factors that bind receptors located on the outside surface of the cell. The signals are then transmitted into the cell and across the cytoplasm to the nucleus. There, by transcriptionally activating the genes for proliferation and by suppressing the genes required for cellular differentiation, the signals are converted into growth responses. As a general rule, oncogenes are mutated forms of the cellular proto-oncogenes which encode the components of this signalling pathway (fig 1).[5] Moreover, the oncogenes are often changed relative to their proto-oncogene forebears in such a way that the proliferative signals are jammed in the "on" position.

Functions of cellular proto-oncogene products

Thus cellular proto-oncogenes can be broadly classified according to the cellular compartment in which their encoded proteins are active. Several proto-oncogenes that encode growth factors

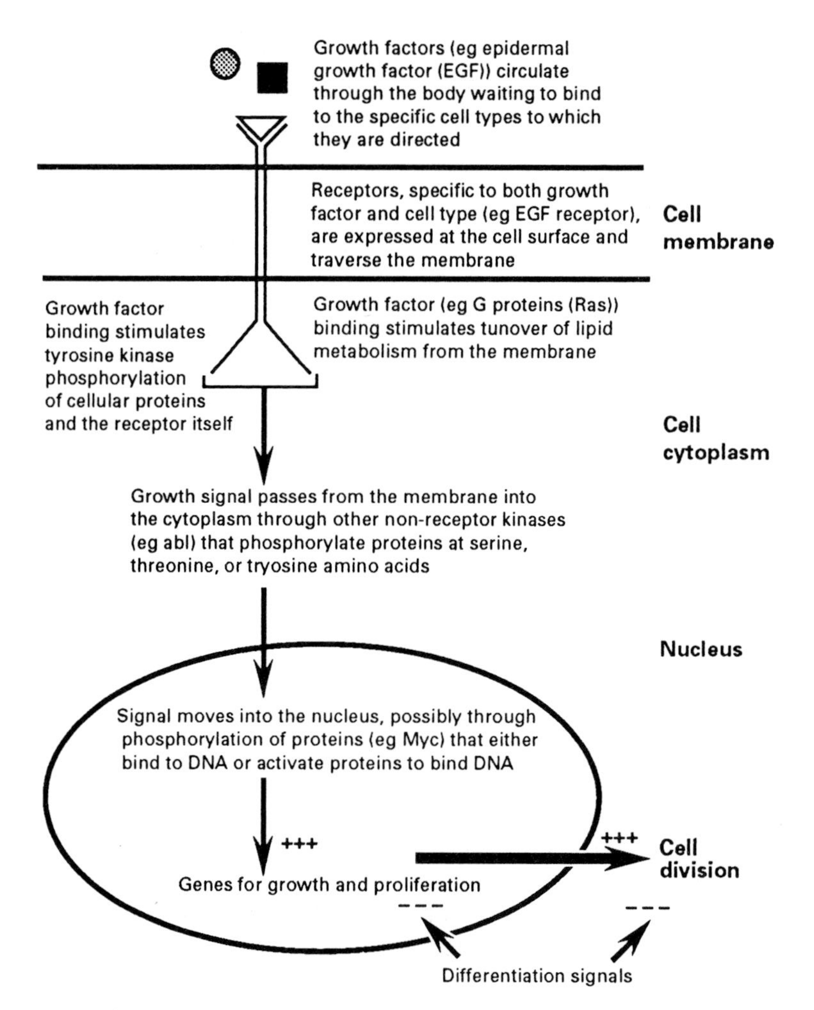

FIG 1—Components of the cell proliferation signal pathway

(such as the platelet derived growth factor (PDGF), c-sis) are known.[6] Clearly, the inappropriate expression of such factors could lead to a continual growth signal being transmitted both to the cell that produces it and to that cell's close neighbours. In some instances the constitutive expression of these growth factors is thought to promote the autocrine growth of the cancer cell, possibly by the growth factor's association with the receptor within the cell (such as in the endoplasmic reticulum). The complex of growth factor and receptor then signals to the cell's own nucleus to maintain the proliferative signal.

Similarly, the products of several proto-oncogenes normally serve as growth factor receptors in the cell membrane (such as epidermal growth factor (EGF) receptor, erb-B). These molecules bind the growth factor and pass the growth signal to intracellular molecules, usually by activating one or more of the second messenger pathways involving inositol lipid turnover and protein kinases A and C. Although the precise signalling pathway is not clear, these receptors often possess tyrosine kinase activity (that is, after binding with growth factor they are capable of phosphorylating certain target molecules on tyrosine residues). Interestingly, the relevant cellular molecules that are the targets for this tyrosine phosphorylation are not known, but in the activated, oncogenic form of the protein this enzymatic activity is usually increased and deregulated, which suggests that it is an important signalling property of growth factor receptors.[7]

Intracellular molecules on the inner side of the cytoplasmic membrane (such as the ras gene family) are then involved in propagating the growth signal, although again the precise biochemical mechanisms are not clear.[8] [8a] Other proto-oncogenes encode proteins, such as c-mos, that operate in the cytoplasm and transmit the signal from the membrane to the nucleus. Some of these proteins also possess kinase activity directed either towards serine, threonine, or tyrosine residues, and deregulation of these activities are thought to contribute to the tumorigenic evolution of the cell.

Finally, many proto-oncogene products are located in the cell nucleus, which is where control of gene expression is principally regulated and so is the ultimate target of the growth signal. These proto-oncogene products (such as c-myc and molecules related to thyroid hormone receptor, like erb-A) are usually involved direct in regulating DNA replication and controlling gene transcription.[9]

In theory, any molecule that is involved in the pathways of cellular transmission of growth activating signals is a potential target for transforming mutations and may eventually be identified as a proto-oncogene. Many new oncogenes have now been identified both from sequences carried by acutely transforming retroviruses and by other techniques, and several are commonly detected in their mutated forms in human cancer cells. The table gives a representative list of some oncogenes associated with transformation, and their grouping by cellular localisation.

TABLE—*Examples (not exhaustive list) of dominantly acting oncogenes that are activated by overexpression or mutation—to illustrate the classes of oncogenes (growth factors, membrane receptors, non-receptor protein kinases, and nuclear oncogenes) that have been identified from tumour cells. (Mechanisms of transformation of potential viral oncogenes by hepatitis B virus (HBV) (hepatocellular carcinoma) and Epstein-Barr virus (EBV) (Burkitts lymphoma and nasopharyngeal carcinoma) not clearly defined.)*

Oncogene	Function of normal protein	Activation by
PDGF B chain (sis)	Growth factor for cells of mesenchymal origin.	Originally isolated from a transforming retrovirus, SSV, fused to viral envelope gene. PDGF is released from many tumour cells, possibly contributing to transformation by autocrine stimulation of growth (megakaryoblastic leukaemia and breast cancer). Activation is generally by over-expression rather than mutation.
hst int-2	Growth factors related to fibroblast growth factor.	hst Was isolated from DNA from human stomach cancer cells and is amplified in some glioblastomas and Kaposi's sarcoma; int-2 is amplified in subset of human breast cancers and glioblastomas. Both proto-oncogenes are angiogenic and may lead to neovascularisation of tumours.
erb-1 erb-2 (neu)	EGF receptor (membrane tyrosine kinase receptor) and receptor kinase similar to EGF receptor (neu).	Transmembrane receptor protein can be activated in vitro by gross deletion (of extracytoplasmic domain) and point mutations, which usually deregulate the tyrosine kinase activity. In human tumours activation is principally by amplification (overexpression): neu is amplified in subsets of breast carcinomas; erb-1 is activated in breast, bladder, brain, and ovarian carcinomas, melanomas, and gliomas.
fms	CSF receptor tyrosine kinase.	All transmembrane receptors with tyrosine kinase activities, which can become deregulated in activated molecules recovered from transformed cells in vitro. Activation is by mutations or deletions: ros is amplified in some breast carcinomas.
kit	Tyrosine kinase similar to PDGF receptor.	
ros	Insulin receptor tyrosine kinase.	
abl	Protein located on inner side of cytoplasmic membrane; has tyrosine kinase activity.	Activated by chromosomal translocation in CGL (see fig 2) as part of Philadelphia chromosome [t(9:22) (q34:q11)] in which tyrosine kinase domain is fused to bcr gene. Biochemical target of normal protein or mutated hybrid not known.

Table continued overleaf

Oncogenes activated by overexpression or mutation (continued)

Oncogene	Function of normal protein	Activation by
pp60-src	Inner membrane, tyrosine kinase protein.	First known retrovirally transduced oncogene. Normal function not known, but tyrosine kinase activity deregulated in squamous cell and stomach carcinomas.
ras	Family of G proteins of unknown function located at inner side of cytoplasmic membrane.	Point mutations to amino acid positions 12, 13, or 61 correlate with activating mutations and the biochemical properties (see text). Activated in various leukaemias, colorectal, stomach, thyroid, bladder, lung, and ovarian carcinomas, neuroblastomas, and hepatomas.
gsp	Inner membrane G protein.	Activated in pituitary cancers.
mil, mos pim	Serine/threonine protein kinases located in the cell cytoplasm.	Activating point mutations increase kinase activities. Relevance to human cancers not yet clear.
myc	Family of nuclear proteins (c-, N-, L-, B-myc). Function not known, possibly affects control of DNA replication or a transcription factor.	Various myc genes amplified in several human tumours (such as breast, cervical, and lung carcinomas), which correlates with poor prognosis. Translocation of c-myc to an Ig locus (chromosome 2, 14, or 22) characteristic of Burkitt's lymphoma. Activation probably by inappropriate levels and timing of expression. High levels of myc genes may be incompatible with process of differentiation.
fos, jun	Nuclear proteins that combine as heterodimers to form the transcription factor, AP-1.	AP-1 regulates expression of genes of differentiation and may control DNA replication. Relevance to human cancers not yet clear.
erb-A	"Zinc finger" containing nuclear protein: a thyroid hormone receptor and transcription factor.	Binds specific DNA sequences in absence of hormone to repress transcription, which is removed when hormone present. Activating mutations (in vitro) remove ability of hormone to bind receptor, so repression of transcription (and differentiation) is not reversible.
myb	Nuclear transcription factor.	Amplified in several human tumours including colorectal and gastric carcinoma.
Viral oncogenes:		
p40 tax	HTLV, transcriptional activator.	Activates transcription of viral and cellular genes (IL 2 and IL 2 receptor and other growth related genes). Associated with ATL.

Oncogenes activated by overexpression or mutation (continued)

Oncogene	Function of normal protein	Activation by
E6, E7	HPV types 16 and 18 nuclear proteins regulating HPV genome expression.	Nuclear proteins that sequester cellular tumour suppressor genes (such as rb and p53) thereby removing critical differentiation signals. Associated with anogenital cancers.
Others:		
CD44	Cell adhesion molecule.	Variant spliced forms can dominantly confer metastatic capacity to non-metastatic cells in an animal model.

PDGF = platelet derived growth factor; SSV = simian sarcoma virus; EGF = epidermal growth factor; CSF = colony stimulating factor; CGL = chronic granulocytic leukaemia; Ig = immunoglobulin; HTLV = human T cell leukaemia/lymphoma virus; IL = interleukin; ATL = adult T cell leukaemia; HPV = human papillomavirus.

Activation of proto-oncogenes to oncogenes

The result of proto-oncogene activation is the increased expression of genes signalling proliferation of cells that should not normally be proliferating. Direct comparisons of the DNA sequences of normal proto-oncogenes with their mutated oncogene derivatives show how proto-oncogenes become activated (that is, converted to the form in which their protein products encode proteins that participate in cell transformation). Activation can occur by changes to the coding sequence of the proto-oncogene so that a mutated protein with aberrant biochemical properties—an oncoprotein—is produced. Such activation events can occur by deletions of portions of the coding sequence (often to remove areas that are required for normal, negative control of the signalling function); by fusion of the sequence to other protein domains that alter the activity of the signalling domain (as occurs in several fusion proteins induced by chromosomal translocation such as the Philadelphia positive translocation in chronic granulocytic leukaemia (CGL), see fig 2); or by point mutations of the sequence at essential bases, which critically alter the amino acid sequence of the protein product. The ras class of proto-oncogenes can become oncogenic by as little as the change of a single base in the sequence of the gene. This alters the protein sequence by only a single amino acid. The Ras protein action is turned off by hydrolysis of

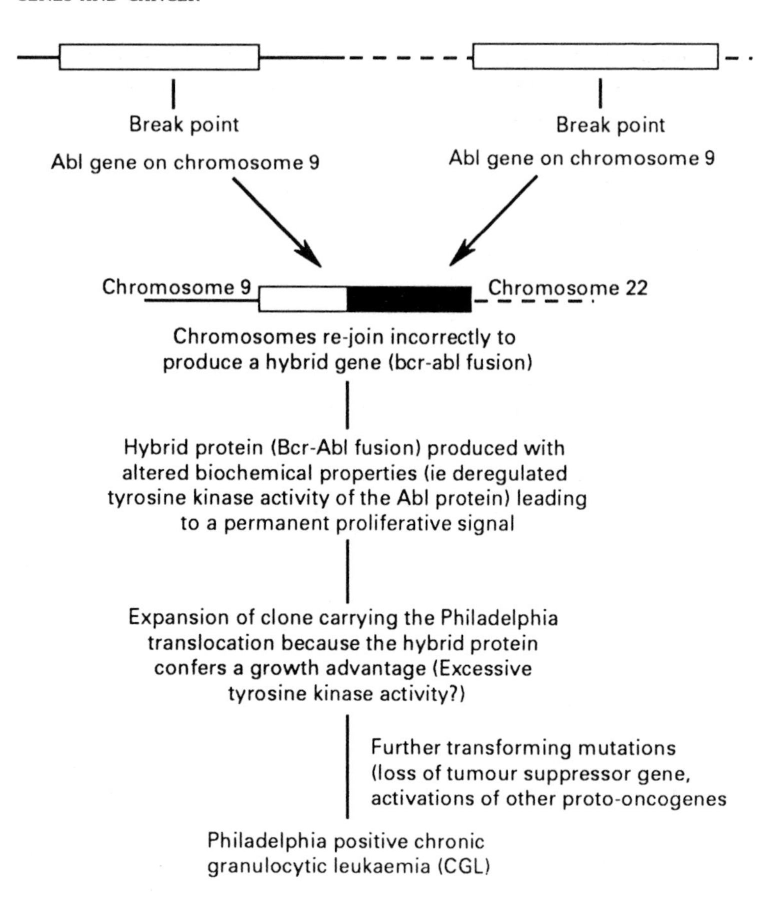

FIG 2—Activation of the abl proto-oncogene by chromosomal translocation (the Philadelphia translocation) to produce a hybrid Bcr-Abl fusion protein

guanosine triphosphate (GTP) to guanosine diphosphate (GDP). The single mutation in the oncogenic form of the Ras protein reduces the ability of the protein to hydrolyse guanosine triphosphate, so the signalling activity of the protein is continually turned on. Up to 15% of human cancers may have this variety of mutation of a ras gene.

Alternatively, activation can occur by mutation that deregulates the levels or timing, or both, of expressing the structurally unaltered proto-oncogene. For example, in the human cancer Burkitt's lymphoma, a chromosomal translocation causes the myc gene to become incorrectly linked to a chromosomal region that permits its permanent expression—even at the incorrect time, when the proliferative signal given to the cell by the Myc protein should be silent. In other human tumours certain proto-oncogenes become amplified, so many DNA copies of the gene exist within the cancer cell, instead of the normal two copies.

Generally it is chemical or physical carcinogens, such as ionising radiation, or mutagenic chemicals that act on DNA, that induce mistakes in the cell's DNA replication machinery or chromosomal instability, which leads to some of the effects on gene structure described above. Infection by viruses, however, may also contribute to the genome instability that characterises the cancer cell and predisposes to genetic damage of this type. Such activation effects can allow proliferative signals to be transmitted, and these mask or compete out the normal signals for differentiation or growth inhibition. The application of molecular biology techniques has now shown that a relatively limited set of cellular genes is repeatedly involved in tumours which are caused by disparate carcinogens.

Interaction of oncogenes

In general, it seems that fully transformed cells must express at least two activated oncogenes, as well as other genetic abnormalities. Many human cancers that have a viral aetiology (about 20%) will also involve interactions of the viral proteins with a subset of these cellular proto-oncogenes. Such is the case with DNA tumour viruses (such as SV40 or papillomaviruses types 16 or 18), whereby the viral "oncogene" (which has not arisen by mutation of a normal cellular gene in these cases) can contribute to the transformation of a cell in much the same way as an activated cellular proto-oncogene. Fully transformed cells tend to express oncogenes that cooperate with each other,[10] and two classes of cooperating oncogenes can be broadly distinguished. At least one of those oncogenes that encode proteins of the membrane receptor or cytoplasmic class is usually expressed in tandem with an oncogene that expresses a protein that is found in the nucleus (table). Expression of a member of either class alone can often produce cellular immortalisation (the ability of a cell to grow indefinitely in culture) but not full tumorigenicity (the ability of a cell to form tumours when transplanted into animals). Each class of oncogene deranges the growth properties of the cell in subtly different ways, and the combination of effects is required for transformation of a cell to full tumorigenicity.

The tumour suppressor genes

Recent discoveries have begun to confirm earlier ideas that mutations in carcinogenesis also involve genes that negatively regulate cell proliferation (and, by inference, promote cell differentiation). Activated oncogenes are generally considered to be "dominantly acting" within the cell because mutation, even in a single copy, appears to be sufficient to drive the cell towards malignancy. However, cancer clearly can no longer be explained solely in terms of activated oncogene expression and mutation, because members of a second class of cellular genes are now emerging as being equally important in driving progression to malignant transformation.

Early cytogenic analysis showed that many human solid tumours show a reproducible loss of genetic material at defined chromosomal loci. These losses are specific for a given tumour type but differ between tumours. The absence of functional genes at these loci seemed to be related to tumour development and, moreover, both alleles (copies on homologous chromosomes) were sometimes seen to be lost (by deletion or gross chromosomal rearrangement). Through observations of the genetic epidemiology of the rare childhood cancer, retinoblastoma, Knudson suggested that a heterozygous child might inherit a (recessive) mutation to one allele of a gene that predisposes to tumours (an inherited first genetic "hit"). Such individuals will go on to develop cancer if the mutation is made homozygous by subsequent damage to the remaining intact allele (the second somatic genetic "hit").[11] This suggestion of a new class of cellular genes that regulate cell growth by counteracting the action of the proto-oncogenes led to the proposal that there are genes that are active at embryogenesis for proliferation ("transforming" genes), which are suppressed at differentiation by dominant "suppressor" genes.[12]

Knudson therefore named this proposed new class of genes "anti-oncogenes" (or "tumour suppressor genes"). The statistical analysis of the occurrence of retinoblastoma in affected families compared with the general population suggested that two "hits" are required, one in each allele of the suppressor gene, before a cell becomes fully predisposed to neoplastic growth. Cancer causing mutations in these genes can therefore be regarded formally as being recessive because both copies of the gene must be defective.

Tumour suppressor genes can be rationalised in terms of the conflicting needs for stem cells both to proliferate and to differentiate. Within a stem cell compartment, cycling cells will only differentiate in response to particular genetic or epigenetic signals; one class of genes (the proto-oncogenes) can simplistically be viewed as the genes required to keep the cells cycling, and the second class (the tumour suppressors) would be those genes that override the proliferative signals and induce the cell to differentiate. The loss of both copies of any of these suppressor genes, before the critical time in a cell's life history at which it is required to interpret the differentiation specific signals, would permit continued, inappropriate expression of the proliferative genes. Such a cell may then be launched on the first stages of neoplastic transformation.

Knudson's concept has been increasingly supported by the findings that several familial cancer syndromes are linked to the inherited loss of a defined chromosomal locus at one allele. This germinal mutation predisposes the individual to the cancer, but for the disease to emerge a second mutation is required somatically at the unaffected allele in the cell(s) that eventually become transformed. In cancers with a sporadic form as well as a familial form, the same chromosomal locus is affected in some cases, but both alleles are inactivated somatically. The differences in ages of patients, and features of the cancers, at presentation support such a model.

There is a rapidly growing list of tumours now being associated with homozygous loss of specific chromosomal loci.[13] Until genetic linkage studies have been carried out for each tumour type, however, it is not possible to know which losses represent germ line alleles important in contributing a genetic predisposition to each specific cancer.

In addition, much in vitro research has implied the existence of genes capable of suppressing malignancy. For example, the fusion of a normal cell with a malignant cell produces a hybrid in which the tumorigenic phenotype is usually suppressed and the differentiation programme of the normal parent cell may be imposed upon the hybrid.[14] Subsequent chromosomal loss from the hybrid can occur, which leads to reappearance of tumorigenicity. The loss of specific chromosomal loci can be reproducibly associated with the reappearance of tumorigenicity, which suggests that tumour suppressive information resides at that locus. These and other experiments, whereby tumour cells can be shown to revert to the non-tumorigenic phenotype by transfer of specific pieces of DNA,[15] permit the identification of some of the proposed recessive loci that normally act to restrain cellular transformation.

The discovery of the tumour suppressor genes promises insights into the processes leading to normal differentiation and the way in which they may be disrupted during cellular transformation. Clinically, it provides a new opportunity to assess cancer susceptibility within families and to screen defined patients for certain types of cancer against which early intervention may be critical. These genes may also lead to new approaches to cancer treatment.

The proteins of some DNA tumour viruses that are known to be involved in causing cancer (particularly SV40 and several papillomaviruses) have been shown to bind to cellular tumour suppressor proteins.[16] Hence the transforming virus proteins may well be acting by inactivating the Rb protein. Infection of a cell before the implementation of the normal differentiation signals therefore removes the effects of the rb gene, which is probably central to the interpretation of those differentiative signals. Differentiation in the relevant compartment of cells is prevented and the infected cell misses its only opportunity to leave the proliferative cycle. It now becomes vulnerable to the action of any oncogenes that have already, or will later, become activated.

Several other tumour suppressor genes have now been identified, such as the p53 gene, the Wilms's tumour (WT) gene, the gene deleted in cancer cells from patients with neurofibromatosis (NF), and the "deleted in colon carcinoma" (DCC) gene.[17] The neurofibromatosis gene has homology with the guanosine triphosphatase (GTPase) activating protein (GAP) which interacts with the Ras class of proto-oncogene proteins, thereby providing a mechanistic link between the tumour suppressors and oncogenes. Loss of tumour suppressor gene expression may be a prerequisite for the development of most cancers, and mutation to the p53 gene may actually represent the most common mutation in all human cancers.[18] For example, a heritable mutation in p53 may be the "first hit" mutation in families with the Li-Fraumeni syndrome, whose members show a greatly increased risk of developing a range of tumours, including breast cancer, osteosarcoma, leukaemia, and soft tissue sarcoma. The inference is that p53 protein may mediate critical cycle decisions in a wide range of tissue types; the absence of the correct protein at certain times during the differentiation programme allows the cells to continue proliferating instead of differentiating.

The genes of metastasis

Although the genes controlling proliferation and differentiation are intuitively the most likely targets for oncogenic mutations, other genetic components of the progression to tumorigenicity are at least as important as the oncogenes and tumour suppressors, at least in terms of clinical progression and outcome of disease. Often, the localised primary tumour cell population itself may pose relatively little direct threat to the patient. Clinically, however, the most life threatening aspect of tumour cell growth is the ability of individual cells to spread from the site of primary growth to distant regions of the body where they can initiate secondary tumours, a process called metastasis. Metastasis is particularly dangerous to the patient because it increases the scope of damage that can be done by the rapidly growing tumour cell population and prevents that damage being restricted to a single site in the body.

Despite the overriding clinical importance of metastasis, the underlying mechanisms are only just becoming tractable to molecular study. Not all tumour cells have the same capacity to metastasise from within the same tumour population. Consequently, the progression from the partially transformed state (essentially a benign tumour mass) to the metastatic state is often considered to involve distinct, but interdependent, events. For tumour cells to become fully metastatic they must be able to acquire several properties that are not present in transformed but non-metastatic cells, and there is now considerable evidence that some or all of these processes are under genetic control.[19] Indeed, whereas it can take many years for a tumour to grow in situ, the conversion to malignancy can often occur in months in animal models, which suggests that relatively few events are required. In non-malignant tumorigenic rat carcinoma cells the only requirement for acquiring a fully metastatic phenotype has been shown to be the expression of a variant of the CD44 cellular adhesion molecule on the cell surface.[20] Similarly, increased expression of the nm23 gene in a melanoma cell line seems to suppress the malignant phenotype of these cells.[21] These studies suggest that the difference between tumorigenic but non-metastatic compared with fully metastatic behaviour may be attributable to the expression of only one or a few genes that promote metastasis or to the loss of an equally small number of genes that exert a suppressive effect on tumour metastasis. In the conversion from normal to tumour to metastatic cell, therefore, there are probably certain rate limiting steps in the overall sequence, which must occur before the malignant phenotype can evolve. For each tumour type these rate limiting steps may differ; and for each metastasising system the individual rate limiting steps may depend on the tumour cell type, its requirements for growth factors, and its ability to escape the host immune response, especially when the tumour cells are exposed in the distributing system (that is, the blood).

The cancer cell and the immune system

Far less well understood, but of great potential importance in treatment, are the genes that control the reaction of cancer cells with the immune systems of patients. As cancer cells develop by a series of genetic and epigenetic mutations of genes within the once normal cells of the body, natural in vivo immune responses to cancer cells would in theory have to be autoimmune, as cancer cells might be expected to constitute an immunologically "self" population that the host's own immune system cannot recognise as foreign and therefore reject. In some circumstances this is assumed to be the case, and tumours can develop within the body unchecked by any natural immune reaction to the aberrant cells.

However, several lines of evidence suggest that cancer cells, far from representing an immunologically hidden population, actually express determinants—tumour antigens—that can form the basis of an anti-tumour immune reaction. Clinical evidence for specific anti-tumour immune activity is seen on the rare occasions when patients experience spontaneous disappearance of cancers, which cannot be attributed to the treatment; and the concept of anti-tumour "immune surveillance" may provide a partial explanation for the observed age related incidence of cancers and for the greatly increased incidence of various forms of cancers in patients whose immunity is suppressed by viral infection or other means. Transplant rejection may represent an in vivo defence mechanism against tumour cells that express antigens that seem to be foreign (due to activating oncogenic mutation).[22]

Largely unsuccessful attempts to vaccinate cancer patients with their own cancer cells are being replaced by treatment protocols aimed at enhancing the immune system's recognition and destruction of emerging solid tumours. These efforts now provide a focus for therapeutic intervention[23 23a] and have stimulated the first in vivo trials for human gene treatment for cancer.[24] Antigen expression can be up regulated by cytokines, and the activity of the antitumour effector cells of the immune system can be increased by treating the patient with lymphokines or preferably by delivering cytokine genes to the site of the tumour, where their expression will promote a vigorous immune response by circulating T cells and monocytes (adoptive immunotherapy).[23] This is the theory behind clinical trials in which the interleukin 2 gene was transferred into tumour infiltrating lymphocytes (TIL), which were reinjected into patients with end stage melanoma. The presence of the interleukin 2 protein should activate these lymphocytes, which presumably have targeting specificity for the tumour (from which they were initially recovered). The early results, admittedly in melanomas that have proved refractory to all other treatments, are encouraging but far from spectacular.[24]

The emergence of a tumour probably represents the outcome of a finely balanced immune surveillance system in which the rate of growth of the transformed cells has finally outstripped the capacity of the immune system to control it. Indeed, some of the critical mutations that finally produce the fully malignant phenotype may be of genes that shift the balance towards evasion of the immune response that otherwise has been successful in clearing, or controlling, the pre-malignant cells.

One specific example of a tumour antigen with immediate clinical importance has been P-glycoprotein (p170), the expression of which is associated with the phenotype giving resistance to many drugs.[25] The expression of this protein in tumour cells has dramatically severe clinical effects because various cytotoxic drugs used to target rapidly growing cells in patients with cancer are bound by p170 and subsequently transported out of the cell. P-glycoprotein is a membrane protein that acts as a pump dependent on ATP for the efflux of drugs and is often responsible for the disappointing recurrence of tumours after initially favourable responses to chemotherapy. In one study 85% of patients with primary, advanced breast carcinoma expressed P-glycoprotein in at least some of the tumour cells.[26] Such studies suggest that monitoring for expression of P-glycoprotein may help predict chemotherapy treatment failure and indicate changing the treatment regimen. Eventually, co-administering P-glycoprotein inhibitors may allow highly effective chemotherapy without the development of the phenotype that gives resistance to many drugs.

From laboratory to clinic

The development of a tumour cell population is a multifactorial process. There is rarely, if ever, a single event to which the conversion of a normal cell to a tumour cell can be solely attributed. Thus, a fully malignant cancer comprises a population of essentially similar cells that harbour many different genetic abnormalities compared with their normal counterparts. It is not clear, however, whether these mutations must occur in a specific temporal sequence.

Diagnosis and prognosis

The genetic changes that occur during cellular carcinogenesis are associated with the cellular acquisition of more aggressive growth characteristics and increasingly malignant behaviour. Specific mutations may, however, be essential for tumour types to emerge in different tissues, the normal differentiation programmes of which are determined by different regulatory genes (for example, mutation of the rb gene seems to be critical to allow cells of the retinal lineage to escape the normal rigours of growth control). Each tumour type may therefore require tissue specific, rate limiting, cellular mutations (tumour initiation) before it can evolve,[27] whereas the nature, timing, and combination of other mutations in the tumour cells may not be crucial (tumour progression).[28]

A main potential benefit of this molecular knowledge will therefore be an increasingly accurate assessment of risk of developing certain types of cancer. Familial cancer syndromes will become amenable to genetic diagnosis using restriction fragment length polymorphisms (RFLPs) of markers linked to the genes that are known to predispose to a given disease. The findings that the loss of certain tumour suppressor genes leads to a high risk of developing cancer will permit the presymptomatic diagnosis of individuals at risk, thereby indicating intensive surveillance and pre-emptive intervention before the cancer develops. Retinoblastoma, Wilms's tumour, neurofibromatosis, and Li-Fraumeni syndrome are some of the cancers the risk of which can now be reliably assessed in members of affected families. Similar studies should identify genes predisposing to more common cancers, such as heritable breast cancers.

A knowledge of the specific oncogene activation events within a cancer cell population is also becoming an increasingly reliable diagnostic and prognostic indicator for disease progression. For example, the Philadelphia translocation can be used as a diagnostic marker for chronic granulocytic disease (this chromosomal translocation is found in about 90% of all patients with this disease). Similarly, patients presenting with acute lymphoblastic leukaemia (ALL) who have t(4:11) karyotypic abnormalities have a very poor clinical prognosis. Conversely, the clinician can now place patients diagnosed as having acute myeloid leukaemia in the best prognostic category if they present with abnormalities of chromosome 16 in the leukaemic cells. In the case of solid tumours, breast cancers that have lost expression of the nm23 gene are strongly associated with very poor patient survival, and would indicate to the clinician a high probability of spread of the cancer cells to the axillary lymph nodes and other widespread metastasis. Finally, knowledge of the "active" oncogenes can be used to detect residual disease in patients—for example, patients with chronic granulocytic leukaemia (CGL) who are in remission can be screened with the polymerase chain reaction (PCR) for leukaemic cells containing the bcr-abl gene. Early detection of relapse by molecular means, before disease has clinically reappeared, will permit better management of patients with recurring disease.

Hopes for specific treatments targeted at cancer cells

The ultimate goal of this wide range of research must therefore be to lead to mechanism based treatments for the disease, and yet many years of research have still not made a large impact on current treatments. The individual stages at which intervention into the molecular pathways of cancer cell proliferation might occur are manifold but also, for that reason, problematic. Pharmaceutical blocking of any of the oncogene products active in a cancer, using rational drug design against the structures (from crystallographic structural studies) known to be altered in the oncoprotein, is therefore attractive; alternatively, it may prove possible to deliver to cancer cells anti-sense DNA or RNA constructs, which block expression of the activated form of an oncogene while leaving the unmutated gene unaffected. In addition, monoclonal antibodies targeted specifically at the mutated regions of an oncoprotein may be able to differentiate between oncogene products and the normal cellular equivalents, which would lead to clearance of cancer cells but not normal cells. Quite apart from the technical difficulties of targeted delivery of these therapeutic agents in the patient, however, by the time the malignant phenotype has emerged the tumour cell may no longer depend on continued expression of any individual mutant protein. Likewise, replacing the functions provided by genes that have been lost in tumour cells—the tumour suppressor genes—has been proposed as a means of reverting to the transformed phenotype (gene therapy). But the loss of such genes may be an ancient event, relative to the growth requirements of the cancer cell, by the time the clinician first sees the patient. Additionally, the interaction of the cancer population with the local and distant environment is central to understanding the life threatening aspects of metastasis. Finally, much poorly understood immunology underlies the control of the emerging cancer cell population by the host immune system. Mutations that lead to loss of the immunological balance between tumour cell destruction and immune escape, and the factors required to restore that balance in favour of tumour cell clearance, are only now beginning to be addressed in molecular terms.

Currently, treatment for cancer can often be as distressing as the disease it seeks to cure. Recent research has shown that primary and secondary tumour populations consist of cells that have undergone multiple mutations, the molecular bases of which are becoming increasingly understood. It is to be hoped that these defects can be exploited by clinicians to target individual, or combinations of, critical cellular mutations, which will allow selective tumour cell killing leaving the remaining normal cells, and therefore the patient, largely untouched by the treatment.

1 Fischer J, Boveri T. *The origin of malignant tumours*. Boveri M, transl. Baltimore: Williams and Wilkins, 1929. Boveri T. *Zur Frage der Erstehung Maligner Tumoren*. 1914.
2 Bishop JM. Molecular themes in oncogenesis. *Cell* 1991;**64**:235–48.
3 Weinberg RA. Oncogenes, antioncogenes and the molecular bases of multistep carcinogenesis. *Cancer Research* 1989;**49**:3713–21.
4 Stehelin D, Varmus HE, Bishop JM, Vogt PK. DNA related to the transforming genes of avian sarcoma viruses is present in normal avian DNA. *Nature* 1976;**260**:170–3.
5 Cantley LC, Auger KR, Carpenter C, Duckworth B, Graziani A, Kapeller R, *et al*. Oncogenes and signal transduction. *Cell* 1991;**64**:281–302.
6 Cross M, Dexter TM. Growth factors in development, transformation and tumorigenesis. *Cell* 1991;**64**:271–80.
7 Ullrich A, Schlessinger J. Signal transduction by receptors with tyrosine kinase activity. *Cell* 1990;**61**:203–12.
8 Bourne HR, Sanders DA, McCormick F. The GTPase superfamily: a conserved switch for diverse cell functions. *Nature* 1990;**348**:125–32.
8a Bourne HR, Sanders DA, McCormick F. The GTPase superfamily: conserved structure molecular mechanism. *Nature* 1990;**349**:117–27.

9 Herrlich P, Ponta H. Nuclear oncogenes convert extracellular stimuli into changes in the genetic programme. *Trends Genet* 1990;**5**:112–16.
10 Hunter T. Cooperation between oncogenes. *Cell* 1991;**64**:249–70.
11 Knudson AG. Mutation and cancer: statistical study of retinoblastoma. *Proc Natl Acad Sci USA* 1971;**68**:820–3.
12 Comings DE. A general theory of carcinogenesis. *Proc Natl Acad Sci USA* 1973;**70**:3324–8.
13 Hollingsworth RE, Lee W-H. Tumour suppressor genes: new prospects for cancer research. *J Natl Cancer Inst* 1991;**83**:91–5.
14 Harris H. The analysis of malignancy by cell fusion: the position in 1988. *Cancer Res* 1988;**48**:3302–6.
15 Huang SH-J, Yee J-K, Shew J-Y, Chen PL, Bookstein R, Friedmann T, *et al.* Suppression of the neoplastic phenotype by replacement of the RB gene in human cancer cells. *Science* 1988;**242**:1563–6.
16 Levine AJ. The p53 protein and its interactions with the oncogene products of the small DNA tumor viruses. *Virology* 1990;**177**:419–26.
17 Marshall CJ. Tumour suppressor genes. *Cell* 1991;**64**:313–26.
18 Hollstein M, Sidransky D, Vogelsttein B, Harris CC. p53 mutations in human cancers. *Science* 1991;**253**:49–53.
19 Liotta LA, Steeg PS, Stetler-Stevenson WG. Cancer metastasis and angiogenesis: an imbalance of positive and negative regulation. *Cell* 1991;**64**:327–36.
20 Gunthert U, Hofman M, Rudy W, Reber S, Zoller M, Haussmann I, *et al.* A new variant of glycoprotein CD44 confers metastatic potential to rat carcinoma cells. *Cell* 1991;**65**:13–24.
21 Leone A, Flatow U, Richter King C, Sandeen MA, Margulies IMK, Liotta LA, *et al.* Reduced tumor incidence, metastatic potential, and cytokine responsiveness of nm23-transfected melanoma cells. *Cell* 1991;**65**:25–35.
22 Thomas L. In: Lawrence HS, ed. *Cellular and humoral aspects of the hypersensitivity states.* London: Cassell, 1959; 529–31.
23 Russell SJ, Lymphokine gene therapy for cancer. *Immunology Today* 1990;**11**:196–200.
23a Russell SJ. Interleukin-2 and T-cell malignancies: an autocrine loop with a twist. *Leukaemia* 1989;**3**:755–7.
24 Rosenberg SA, Abersold P, Cornetta A, Kasid, A, Morgan RA, Moen, R, *et al.* Gene transfer into humans—immunotherapy of patients with advanced melanoma, using tumour infiltrating lymphocytes modified by retroviral gene transduction. *N Engl J Med* 1990;**323**:570–8.
25 Endicott JA, Ling V. The biochemistry of P-glycoprotein-mediated multidrug resistance. *Annu Rev Biochem* 1989;**58**:137–71.
26 Verrelle P, Meissonnier F, Fonck Y, Feiuel V, Dioneto G, Kwiatowski F, *et al.* Clinical relevance of immunohistochemical detection of multidrug resistance P-glycoprotein in breast carcinoma. *J Natl Cancer Inst* 1991;**83**:111–6.
27 Haber DA, Housman DE. Rate-limiting steps: the genetics of pediatric cancers. *Cell* 1991;**64**:5–8.
28 Stanbridge EJ. Identifying tumor suppressor genes in human colorectal cancer. *Science* 1990;**247**:12–3.

Molecular immunology of antibodies and T cell receptors

Richard G Vile

Introduction

The immune system has two arms, which provide specific immunity or non-specific immunity. Non-specific immune mechanisms continually protect the body against invasion by a wide variety of infectious agents with an unchanging response, irrespective of the nature of the agent or whether the body has been exposed to it previously. Specific immunity is acquired against only one particular foreign substance (an antigen) at a time. Moreover, it generates the lasting advantage of immunological memory; the next time the agent is encountered the response against it is rapidly mounted again and amplified. Without this property, vaccination would not be possible.

The effector cells of specific immunity are called lymphocytes. Although morphologically very similar, they are grouped into functionally distinct subsets, the broadest of which distinguishes B lymphocytes from T lymphocytes. B cells develop in the fetal liver and the adult bone marrow, whereas T cells mature in the thymus. Specific recognition of antigens in the immune system occurs by specialised receptors bound to or secreted by cell membrane, which are produced by the B cells (antibodies) or the T cells (T cell receptors (TCRs)). Communicating with the antibodies and the T cell receptors are several other proteins, which allow a biochemical signal to be passed to the nucleus of the lymphocyte after antigen recognition. Depending on the specific type of lymphocyte, this signal produces cellular activation and the starting of the immune response.

Humoral immune response—antibody secretion by B cells

Immunoglobulin gene structure

B cells are distinguished by the synthesis of antibodies and, in some cases, by their presence on the cell surface. Antibody production by B cells generates the humoral immune response.

Antibodies are composed of immunoglobulin polypeptide chains encoded by heavy and light chain genes. Each antibody is composed of two identical heavy and two identical light chains. However, no single gene encodes the complete heavy or light chain polypeptide itself in immature B cells. Instead, separate gene segments exist that contribute to the final heavy or light chain gene, which is expressed exclusively in any one B cell. Similar gene segments are grouped together in clusters on the chromosome, and the clusters of segments of different types are separated by wide distances, but contiguously, along the same chromosome. In the developing B cell, one segment from each cluster undergoes gene rearrangement. The rearranged segments come together to form a unique combination, which is now a functional gene encoding a complete heavy or light chain polypeptide. The different combinatorial rearrangement of these gene segments in each immature B cell creates a set of related, but distinct, immunoglobulin genes and allows different B cells to express immunoglobulins with different structures. Each structure can then potentially recognise a different antigen, which accounts for the enormous diversity of antigens that can be neutralised by an individual's own antibodies.

Immunoglobulin light (L) chain genes are assembled by the rearrangement of three segments: the variable region (V), the joining region (J), and the constant region (C). Immunoglobulin heavy (H) chain genes are also formed from V, J, and C regions but have a further region called the diversity region (D) (see fig 1). In a "typical" antibody, two identical heavy chains fold together with two identical light chains (see fig 2). The H and L chains are joined by interchain disulphide bonds, whereas intrachain disulphide bonds help to form a distinctive protein folded structure, which recurs in both types of chain, known as the immunoglobulin fold or immunoglobulin domain (shown as semicircles in fig 2). The V, J, and D regions of the H chain (denoted V_H in fig 2) and the V and J regions of the L chain (V_L) come together in the final folded antibody molecule to form the site at which foreign antigen is bound (the antibody combining site). Different combinations of V,

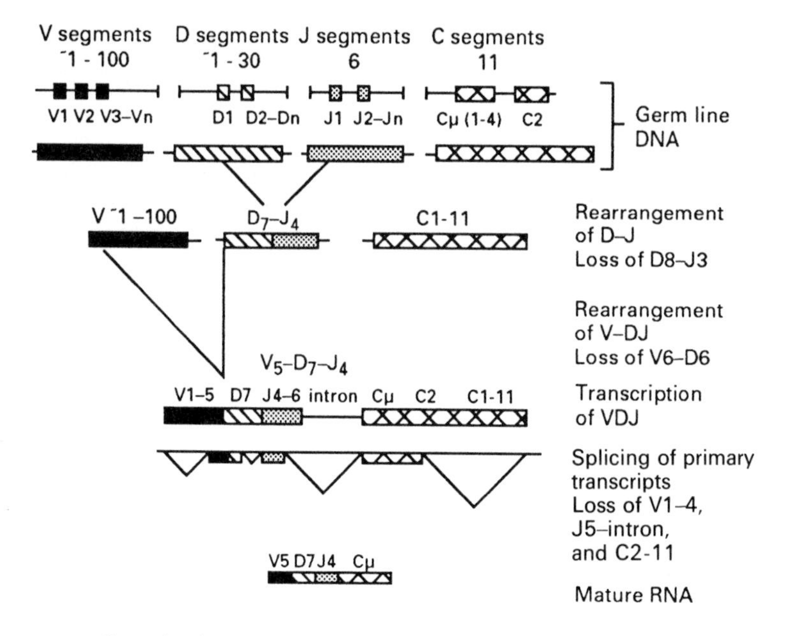

FIG 1—Organisation, rearrangement, and transcription of the heavy chain gene segments of human immunoglobulin. The contiguous clusters of V (variable) segments (about 100), D (diversity) segments (about 30), J (joining) segments (6), and C (constant) segments (11) are shown. Only Cμ of the C segments is shown, and this encodes four immunoglobulin domains (see fig 2).

The first DNA rearrangement joins a D segment to a J segment; the second brings a V segment to the rearranged DJ unit. At the RNA stage the remaining introns and other segments are moved from the primary transcript of the rearranged DNA to give the mature RNA, which encodes the heavy chain protein.

Similar rearrangements at the light chain locus produce a mature light chain protein. Two heavy chains and two light chains then fold together to produce the functional immunoglobulin.

D, and J regions in the final H and L genes therefore determine the antigen specificity of the B cell antibody.

The κ light chain gene segments are located on chromosome 2, the λ light chain gene segments are found on chromosome 22, and the heavy chain gene segments are on chromosome 14. The κ locus consists of 100–300 V segments, 5 J segments, and 1 C segment; the λ locus has several V segments, 6 J, and 6 C segments. The heavy chain locus consists of more than 100 V, 30 D, 6 J, and 11 C segments (fig 1). Each C segment of the heavy chain locus encodes either 3 or 4 immunoglobulin domains. These regions of the

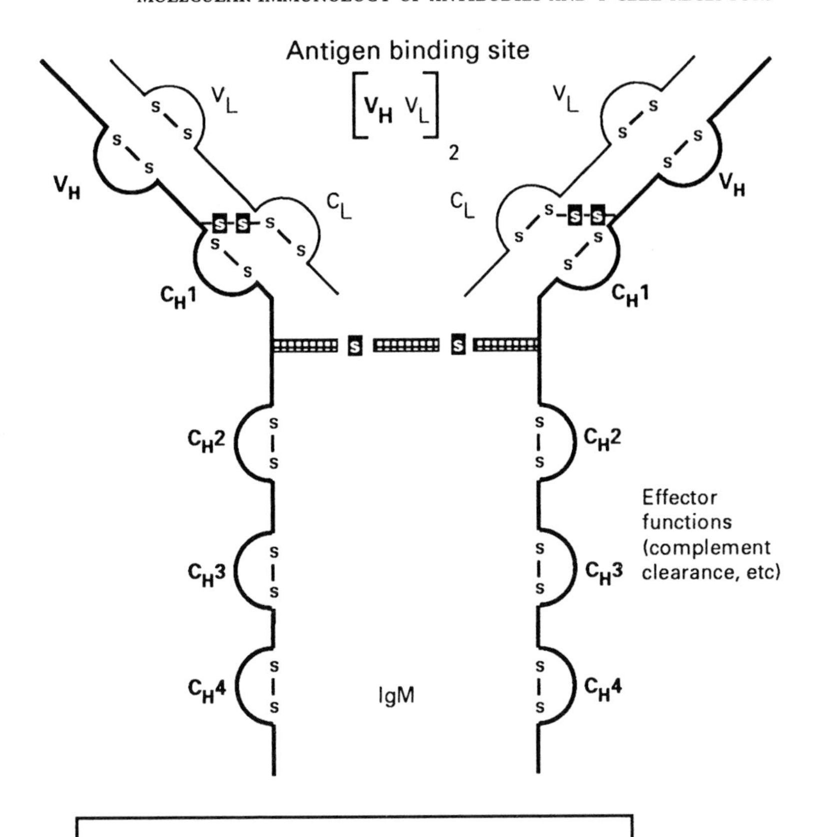

FIG 2—Protein structure of an immunoglobulin molecule showing the correlation of gene segments to structural domains. In this example the heavy chains have four constant region domains (immunoglobulin domains) all encoded by the Cμ segment.

molecule determine the effector functions of the antibody (such as the method by which antigen complexed with antibody is removed) and are not subject to variability in structure. They are encoded by "constant" C gene segments; the different heavy chain C segments determine to which class of immunoglobulin molecules the final antibody belongs (see below).

Immunoglobulin gene rearrangements

The individual V, D, and J segments are flanked by functional units of non-coding DNA. These consist of either 7 or 9 base pair sequences separated by spacers of either 12 or 23 base pairs. A V segment may rearrange and join only with a D or J segment that has the alternative type of spacer to its own (the 12 or 23 base pair rule), which is probably a structural restriction imposed by the endonucleases and ligase proteins that mediate V-D-J rearrangements.

In the immature B cells the immunoglobulin loci on chromosomes 2, 14, and 22 each consist of clusters of V, D, J, and C segments in a linear sequence on the respective chromosome (fig 1). Few RNA transcripts are produced from any of the individual segments, and any that are do not produce protein. During B cell maturation, rearrangement occurs at the immunoglobulin loci. For example, at the heavy chain locus, the 3′ end of a D segment is first brought into juxtaposition with the 5′ end of a J segment (see fig 1). The intervening DNA, which separated these segments in the genomic DNA and contained other D and J segments, is usually looped out and lost. Then the 3′ end of a V segment is brought close to the 5′ end of a DJ segment, removing the intervening DNA. The rearrangement of the genomic DNA segments brings the promoter regions (5′ of the V segment) of the immunoglobulin gene into close proximity with the immunoglobulin enhancer, which lies between the J and C segments. This spatial association of the DNA control elements, which promote active gene expression, activates high level transcription of the rearranged heavy chain gene. The resulting VDJ unit then becomes transcriptionally active and RNA transcripts are produced that include a C region segment. These transcripts undergo RNA splicing to remove RNA derived from any remaining intervening sequences. The final mature RNA encodes the heavy chain gene that will be exclusively expressed by the B cell. Light chain rearrangements concurrently occur at either the κ or the λ locus so that the B cell eventually synthesises one type of heavy and one type of light chain only. Association of the two chains after protein translation gives an antibody with a highly specific antibody combining site formed by the unique association of VDJ and VJ rearrangements (fig 2).

Antibody diversity is generated by the combination of rearrangements of immunoglobulin segments

The immune system must be able to provide an effective humoral response to a huge variety of antigens, which cannot be predicted. To encode all the potentially necessary antibody combining sites as "hard copies" in the genome would involve, theoretically, an infinite number of predetermined genes. This is clearly impractical and would be tremendously wasteful. Instead, the DNA that is encoded at the immunoglobulin loci is used in a more efficient way. Random rearrangements make it possible to generate a repertoire of antibody combining sites that is likely to recognise most antigens that will be encountered during life.

For example, at the heavy chain locus, more than 100 V, 30 D, and 6 J segments exist. The provision of different hard copies of each segment—the germ line encoded diversity—provides a first level of variability. Any V segment can join to any D and any D to any J (within the limitations of the 12 or 23 base pair rule). Therefore, a total of 18 000 (100 × 30 × 6) VDJ combinations are possible in the heavy chain alone.

The germ line encoded diversity is further amplified by built-in randomness within the rearranging process. D-J and V-DJ joining does not always occur at the same place within each segment so that different lengths of D segments are produced. Therefore, imprecise joining of the same germ line V-D-J segments could produce series of different VDJ sequences.

Further diversity is introduced by the activity of the enzyme terminal deoxyribonucleotidyl transferase, which randomly inserts nucleotides into the DNA at the segment joining points (sometimes called the N region of the final gene). VDJ combinations can be produced, therefore, which have elements of DNA that were never actually encoded by the genome. The additional variety produced by these imprecise mechanisms is difficult to assess but may add a factor of 10 to the 18 000 VDJ combinations produced by the mixing of the germ line encoded segments.

Similar calculations can be made for the diversity generated from the κ and λ loci for the light chains. Any light chain can probably pair with any heavy chain, so this combinatorial diversity produced by the pairing of heavy and light chains generates a fourth layer of complexity in the generation of the repertoire. Using these four mechanisms (germ line encoded diversity of V, D, and J segments; imprecise joining of rearranging segments;

random base pair insertions by terminal deoxyribonucleotidyl transferase; and combinatorial diversity of heavy and light chain pairing) more than 10^9 antibody combining sites could be generated with the genetic information that is encoded at the three immunoglobulin loci.

Finally, antibodies synthesised by mature B cells can also accumulate mutations in the variable regions after they have encountered antigen (see below). This process, known as somatic mutation, can further add to the repertoire of potential specificities of antibodies or it can be used to alter the affinity of an antibody for its specific antigen.

Generating the antibody response

The first step of the antibody response is the exposure of the immune system to a foreign (non-self) antigen. The antigen is engulfed by cells of the reticuloendothelial system, including macrophages and dendritic cells, which are classed as antigen presenting cells (APCs). This uptake is largely non-specific, and the antigen is partially degraded within a lysosome in the antigen presenting cells. Fragments of eight or nine amino acids are then circulated to the cell surface again, bound to a groove at the surface of a heterodimeric class II major histocompatibility complex (MHC) protein. The antigen fragment is thereby presented at the cell surface where it is "seen" and bound by a helper T lymphocyte. The T cell bears a T cell receptor on its surface (see below), which specifically recognises the antigen fragment but only in association with the class II major histocompatibility complex molecule. The T cell will recognise correctly presented antigen only, not just vacant major histocompatiblity complex or unprocessed antigen.

Binding of the T cell receptor by the antigen stimulates the T cell to proliferate and to differentiate. This requires local secretion of interleukin 1 by the antigen presenting cells. Thus T cells that are bound to antigen presenting cells and can recognise the correct antigen are the only cells stimulated to divide. The T cell also expresses receptors for other growth factors, most notably interleukin 2. In response to interleukin 1 and antigen binding, T cells begin now to secrete interleukin 2. The T cell proliferation that results from interleukin 2 binding permits rapid amplification of the helper T cell response, but only for T cells specific for the original antigen.

Meanwhile, resting B cells, in which the immunoglobulin genes are rearranged and which are expressing specific antibodies on their surface, are also exposed to the antigen. Only one, or a very few, B cells will express antibody, the combining site of which is complementary to the antigen; these B cells will bind the antigen and internalise it, where it will again be partially degraded and re-expressed in association with class II major histocompatibility complex molecules on the surface of the B cell. The helper T cells, which have already been primed for the antigen by presentation to antigen presenting cells, will now recognise the class II associated antigen fragment on the B cell surface and will bind to it by the T cell receptor. The binding of activated helper T cells to B cells provides the essential stimulus for B cell proliferation and differentiation. The helper T cells secrete interleukins 4 and 5, which induce the differentiation of the activated B cells into plasma cells and memory cells. Plasma cells are short lived B cells that secrete large amounts of the specific antibody, which are usually sufficient to bind to, and clear, the antigen against which the entire cascade is directed. A few of the B cells differentiate into memory B cells. These are very long lived in the circulation and express membrane bound antibody with the specificity for the antigen. These cells, with the long lived helper T cells, are "antigen primed" so that when the antigen is encountered in the future they can start a rapid and intense humoral response to clear it. The existence of pre-existing primed T and B cells in the circulation explains the greater vigour of the secondary response to an antigen and is the basis of vaccination programmes.

During the differentiation of the immature B cell in response to antigen binding and T cell "help", immunoglobulin class shifting occurs to alter the nature and affinity (but not the specificity) of the antibody produced by the B cells. The resting B cell will express surface bound antibody with heavy chains of the μ class. In these cells a VDJ unit is joined to a Cμ segment (figs 1 and 2). Five classes of heavy chains, which differ in the C segment to which the VDJ unit is joined, exist in mammals. Antibody containing Cμ heavy chain is called IgM, is membrane bound, and is the predominant class of antibody in the primary response. Later in B cell differentiation, however, the same functional VDJ unit (antigen specificity) can be expressed with a different C segment by alternative splicing patterns (fig 1). The C segment of an antibody determines its properties other than its antigen specificity, and determines the immunoglobulin class. Typically, IgM is replaced during B cell maturation by IgG antibodies in which the Cγ segment is expressed with the VDJ unit. IgG is the predominant immunoglobulin in the secondary response, and the Cγ region of the heavy chain allows antibody complexed with its antigen to be removed from the circulation by complement fixation. The other immunoglobulin classes, IgA, IgD, and IgE, all have distinctive effector functions determined by the C segment of the heavy chain.

Antibody maturation also occurs during differentiation of B cells. Somatic mutation (see above) of the variable regions of the H and L chain genes expressed in the maturing B cell generates a pool of B cells expressing antibodies with slightly different affinities for the same antigen. B cells that express the highest affinity antibodies will be slected because they bind most strongly to the antigen. These B cells will go on to differentiate into plasma and memory cells, thus providing the best "quality" antibodies to combat the antigen.

Cell mediated immune response and T cell receptor

T lymphocytes migrate from the bone marrow to the thymus to complete their development. There are two principal subsets, helper T cells and effector T cells. Helper cells modulate their own activity as well as that of other cells (such as B cells in generating an antibody response); effector T cells, such as cytotoxic T lymphocytes (CTL) and delayed type hypersensitivity T cells, mediate the selective killing of cells seen as foreign.

On exposure to a foreign antigen T cells are stimulated to express either the helper or the cytotoxic T cell phenotype. This T cell activation is a direct result of the interaction of the T cell receptor complex on the surface of the T cell with an antigen (fragment) bound to major histocompatibility complex proteins at the surface of an antigen presenting cell (see above). The T cell receptor shares many similarities with antibody in structure and composition, including component polypeptide chains consisting of repeated immunoglobulin domains; the most important difference is that antibody has evolved to recognise antigen alone but the T cell receptor requires antigen to be presented in association with major histocompatibility complex.

The ligand binding domain of the T cell receptor is contributed by a heterodimer of either an α and β chain or of a γ and δ chain. These subunits are highly polymorphic and recognise the unique antigen fragments in association with major histocompatibility complex. They represent the direct analogues of the heavy and light chains of immunoglobulins. Each chain has a constant and a variable region and is encoded by a composite gene, which arises from recombination of individual segments. However, in addition to the αβ or γδ chains, the membrane bound T cell receptor is closely associated with five other chains referred to as CD3-ζ. These components are required for efficient surface assembly of the receptor and for transmission of the biochemical signals for T cell activation to the nucleus on binding of antigen.

T cell receptor structure

Most peripheral T cells express the disulphide linked αβ T cell receptor, but about 5% of T cells express γδ heterodimers. There are single α, β, and γ chain loci, but three different types of δ chain gene loci have been described. The functional differences of the different receptor types are unknown, although the proportion of cells expressing the αβ form increases from ontogeny through adulthood.

The α gene locus, on chromosome 14, is analogous to the light chain genes of immunoglobulins. There are three clusters of segments, encoding V, J, and C segments. The β chain locus on chromosome 7 is composed of four segments, V, D, J, and C, thus being analogous to the heavy chain of immunoglobulins. Despite several differences in the organisation of the gene loci compared with the immunoglobulin loci, rearrangement of the gene segments occurs to generate functional α and β genes. Gene recombination for the β segments also follows the 12 or 23 base pair rule. Rearrangement of the γ and δ gene loci is also similar, with any one of multiple gene segments in each cluster being able to contribute to the final functional gene. (The δ locus actually lies between the V and J segment clusters of the α locus and is removed by rearrangement of the α segments. A T cell with a rearranged α gene therefore no longer has the genetic information to encode a δ gene.) Similar mechanisms therefore exist for generating mature T cell receptors and immunoglobulin genes. Given the structural similarities of the polypeptide products, the T cell receptors and antibody proteins are probably two evolutionary products derived from one common ancestor, built to a common design but with very distinct functional properties.

The same diversity that is required of antibody structure is similarly required to produce sufficient T cell receptors to recognise all the antigens that might be encountered (see above). This diversity arises exactly as it does in the production of immunoglobulins—by gene rearrangements between germ line derived diversity of gene segments, imprecise joining of those segments, and random nucleotide insertions to generate N regions. However, somatic mutation does not seem to occur in the generation of T cell receptors as it does during antibody maturation.

T cell receptor assembly and signalling

The polymorphic αβ or γδ chains are non-covalently associated with a set of invariant molecules in the CD3-ζ complex. The CD3γ, CD3δ, and CD3ε subunits are closely related and are encoded by linked genes on chromosome 11. The CD3ζ and CD3η molecules are produced by alternative splicing of RNA from a single gene on chromosome 1. These two polypeptides are linked by disulphide to each other and to themselves in the mature T cell receptor complex. The role of the CD3 subunits seems to be in transducing the T cell activation signal to the nucleus after binding of antigen by the αβ chains. Two signal transduction pathways are known to be triggered on stimulation of the T cell receptor by antigen. The first involves activation of a phospholipase C enzyme specific for phosphatidylinositol. This leads to release of inositol triphosphate (ITP) and diacylglycerol (DAG), which mobilize intracellular stores of Ca^{2+} ions and protein kinase C respectively, and these act as "second messengers" of signal transduction. Within seconds of T cell receptor stimulation several proteins also become phosphorylated on tyrosine residues, which suggests that a tyrosine kinase is activated. A member of the Src tyrosine kinase family of proteins, Fyn, has recently been shown to coprecipitate with CD3. CD3 and the $p56^{lck}$ kinase are excellent candidates for the T cell receptor activated kinase activities. The intermediates that link the T cell receptors with the kinases remain to be discovered, however, although the ζ subunit of the T cell receptor has been shown to be sufficient to couple stimulation of the T cell receptor with the tyrosine phosphorylation pathway.

The CD3 subunits probably also regulate the surface expression of the T cell receptor because incompletely assembled complexes are rapidly degraded. In particular, synthesis of the ζ chain is rate limiting whereas the other components of CD3 are synthesised in vast excess.

By analogy, the B cell receptor (immunoglobulin) also appears to be non-covalently associated with other molecules when it exists as a membrane bound form (that is, IgM). Evidence suggests that immunoglobulin is associated with an $\alpha\beta$ dimer (not related to the $\alpha\beta$ of the T cell receptor) and that the mb-1 gene, which encodes the α subunit, has homology with some of the CD3 proteins.

Generating the cell mediated immune response

T cell receptor recognition of antigen (in association with class II major histocompatibility complex) on helper T cells has already been discussed (see above). Cytotoxic T cells are important in immunity against intracellular invading pathogens, such as viruses, and in recognising potential tumour antigens expressed preferentially on the surface of tumour cells. These T cells recognise antigen fragments bound to class I major histocompatibility complex molecules on the surface of the infected cell by their T cell receptors. In this case the antigen presenting cell is the infected or tumour cell, and endogenously synthesised antigen fragments are presented in the surface cleft of the major histocompatibility complex molecule.

The class I molecule, with complexed antigen, is bound specifically by the T cell receptor and CD3 complex. Binding depends on the ability of the antigen binding (variable) region of the T cell receptor $\alpha\beta$ heterodimer to recognise the antigen. This is directly analogous to recognition of antigen by antibody through its variable region, which is contributed by the heavy (VDJ) and light chain (VJ) polypeptides.

Several adhesion molecules are required simultaneously on the T cell to increase the stability of the interaction of major histocompatibility complex and antigen with T cell receptor and CD3. For example, the CD4 molecule on helper T cells binds specifically to the non-polymorphic determinants of class II major histocompatibility complex molecules; the CD8 molecule on the cytotoxic T lymphocyte binds to class I major histocompatibility complex. Other molecules are also required to optimise interactions between the T cell and the antigen presenting cell. After the intimate membrane to membrane interaction of T cell and antigen presenting cell, intracellular signals are started by the T cell receptor and CD3 complex, which lead to changes in gene expression and produce T cell proliferation and differentiation (see above). Among the changes that result are expression of cytokines and their receptors. Interleukins 1, 2, 4, 6 and 7 influence the proliferation or differentiation of helper T cells (restricted by class II major histocompatibility complex) whereas interleukins 2, 4, 6, and 7 and interferon gamma direct the maturation of the cytotoxic T lymphocyte population. It is not clear whether intimate contact between cytotoxic T lymphocytes and helper T cells (CD4+) is required to activate cytotoxic T lymphocytes. Helper T cells specific for a given antigen are certainly required to produce the activating cytokines for cytotoxic T lymphocyte proliferation. Some have argued that both helper T cells and cytotoxic T lymphocytes are simultaneously activated through a ternary cellular complex between helper T cell, cytotoxic T lymphocyte, and antigen presenting cell. This would achieve the specifically antigen induced local secretion of cytokines required for cytotoxic T lymphocyte expansion.

After full T cell activation, cytotoxic T lymphocytes are capable of lysing the cells expressing the antigenic determinant. Activated cytotoxic T lymphocytes secrete degradative enzymes (for example, serine esterases), tumour necrosis factor, and channel forming perforins, which are inserted into the target cell membrane. The resulting pores lead to rapid lysis of the target cell (in this case the antigen presenting cell) without damage to the cytotoxic T lymphocyte. How the cytotoxic T lymphocyte protects itself from its own secreted perforins and lysis inducing molecules is not known, although an inhibitor of perforin mediated lysis has been found in granules in cytotoxic T lymphocytes.

Just as memory B and helper T cells persist after initial clearance of antigen, so memory cytotoxic T lymphocytes remain in the circulation and permit a rapid and intense response to any future appearance of similarly infected cells. These cells retain specificity for the antigen by continued expression of the T cell receptor with complementarity to the antigen. Memory T cells, as opposed to immature or virgin T cells, are characterised by low levels of expression of the CD45RA membrane marker.

The molecular pathology of antibody and T cell receptors

The remarkable ability of the immune system to clear most infections reflects the efficiency with which sufficient antibody and T cell receptor diversity is generated in vivo. However, there are instances in which the defence afforded by the humoral and cellular immune systems can be ineffective, or even subverted, resulting directly in disease.

For example, many patients who are infected with the human immunodeficiency (retro) virus (HIV) appear to produce very good antibody responses to the virus. Indeed, the presence of circulating antibody is the principal diagnostic marker for infection. Some of the antibodies from patients are even capable of neutralising the infectivity of the virus in vitro, and yet such patients still go on to develop AIDS. The mechanisms of how HIV causes disease are not yet understood, but clearly the production of antibodies alone is insufficient. Perhaps no antibody is produced with the specificity to recognise the most appropriate part of the virus that would lead to total neutralisation in vivo. This may be because HIV produces antigens that are masked from the immune system. This can occur when a fragment of an antigen does not bind to a class II molecule and so cannot be presented to helper T cells by antigen presenting cells or B cells. In such cases no B or T cell response can be mounted because no T cell help is available for B and T cell activation.

An effective cytotoxic T lymphocyte response is probably also especially important for retroviral infections because retroviruses can lie latent in infected cells for long periods. When the virus is activated the infected cell acts as a factory for chronic production of virus. The only way to clear infection is to destroy the infected cells by a cytotoxic T lymphocyte response to viral antigens on the surface of the infected cells. Exactly which, if any, viral proteins form the most effective target antigens for an anti-HIV cytotoxic T lymphocyte response is not yet known. These antigens would be excellent candidate components for a vaccine against HIV. A further complication is that HIV infects and can kill helper T cells, which are central to both the humoral and cell-mediated responses against infections of all kinds.

A more recent hypothesis has suggested that AIDS might be caused by an antibody response to the virus which becomes misdirected. A structural similarity has been detected between the viral envelope protein and class II major histocompatibility complex molecules, so antibodies raised in the normal way to the viral antigen might cross react with the class II protein. Class II major histocompatibility complex molecules are central to the generation of efficient antibody responses to infection (see above). Antibodies against these molecules, originally intended to attack HIV, could possibly disrupt the normal functioning of the immune system. As AIDS is manifested as a breakdown of resistance to infection, loss of major histocompatibility complex class II function of antigen presenting cells could explain many of the clinical features of the disease, although this remains to be proved.

Other autoimmune diseases can also be attributed to a derangement of normal immune responses in which they are turned on the host rather than an infectious agent, whether or not an infectious agent (for example, HIV), is present. Multiple sclerosis is an example in which a pathogenic role has been proposed for the particular genes of the T cell receptors, which are rearranged in some T cells. Susceptibility to disease has been linked to the precise V segment genes that are used in the T cell receptors of patients with multiple sclerosis. Although much conflicting data have been published, several studies have shown preferential usage of specific V genes in the autoreactive T cells of patients with multiple sclerosis compared with those of controls. As multiple sclerosis is characterised by degradation of the myelin sheath around nerve cells, T cells reactive against myelin antigens are possibly generated, maybe after infection by another infectious agent. The T cell receptor that is complementary to the infectious agent's antigen might also recognise myelin antigens, and such T cells would begin to attack the nerves of the host. Only certain V segments of the T cell receptor α and β chains would have the complementarity of the antigen binding site, and these could also recognise myelin proteins. Only people in whom these particular genes are rearranged to form functional T cell receptor would be susceptible to multiple sclerosis. The nature of the infectious agent against which the potentially autoreactive T cells are activated remains unknown. In both AIDS and multiple sclerosis the role of autoreactive antibody and T cell receptor remains to be proved; however, other diseases may also exist (such as rheumatoid arthritis) in which an antibody or T cell response is originally mounted against a genuine pathogen but becomes perverted into a pathological autoimmune reaction.

The mechanisms used to create antibody or T cell receptor diversity involve highly complex molecular rearrangements at the DNA level and are unique to T and B cells. It seems inevitable that such specialised reactions should go wrong occasionally. This manifests itself in several B and T cell cancers that arise direct from active gene rearrangements occurring at the immunoglobulin and T cell receptor gene loci. The enzyme controlled rearrangements within molecules between V, D, and J segments sometimes become rearrangements between molecules when the immunoglobulin or T cell receptor locus becomes joined not to another part of the same chromosome but instead to another chromosome by mistake. This chromosomal translocation results in the immunoglobulin or T cell receptor locus becoming juxtaposed to another gene. If this gene is a proto-oncogene (see previous chapter) its expresion can become deregulated and the B or T cell is pushed one step towards malignant transformation. This occurs in Burkitt's lymphoma, a B cell malignancy principally of children and young adults, in which the c-myc proto-oncogene (chromosome 8) becomes joined to an immunoglobulin locus on chromosome 2, 14, or 22. The juxtaposition of c-myc, which is intimately involved in regulating the rate and frequency of cell division, to the immunoglobulin locus, in which a strong, constitutively active enhancer now promotes its overexpression, is a critical step in the transformation of the B cell (see previous chapter). The need for the immunoglobulin or T cell receptor genes to rearrange in B and T cells makes such molecular mistakes possible, and chromosomal translocation break points involving both immunoglobulin and T cell receptor loci are seen in other B cell (for example, follicular lymphoma) and T cell (acute and chronic forms of T cell leukaemia) tumours.

Conclusion

The diversity of antigen recognition by the immune system is expressed in the repertoire of combining sites of antibodies and T cell receptor molecules. Although the repertoire is clearly sufficient to neutralise most of the antigens that are encountered in a lifetime, there are gaps through which some antigens can slip. The ability of the antibodies and T cell receptors to distinguish between "self" (not to be attacked) and "non-self" (to be attacked) antigens stems from the recognition of antigen in association with major histocompatibility complex molecules by the T cell receptors on T cells. In most people, this process (self tolerance) works remarkably well by mechanisms that are not clearly understood. However, the selection system is not foolproof, and it seems that some non-self antigens can generate a B or T cell response that coincidentally recognises self antigens and leads to autoimmune syndromes. Other autoimmune diseases may represent a breakdown of the self versus non-self toleration mechanisms by which autoreactive T cells are normally deleted in the thymus. In these instances, a foreign antigen may not be necessary to induce autoimmune reactions. Finally, the molecular dynamics of DNA structure rearrangement allows complete genes to be assembled from small segments and permits the generation of antibody and T cell receptor diversity. Built into these mechanisms is a degree of imprecision. Within certain limits this randomness (that is, combinatorial joining of multiple gene segments, imprecise joining) is essential to amplify the built-in diversity encoded by the genome. However, when these limits become exceeded (completely abberant joining of different chromosomal segments) molecular mistakes can occur, which can lead to the induction of B or T cell tumours.

Further reading

Clark WR. The basic biology of T and B cells. In: *The experimental foundations of modern immunology*. New York: Wiley, 1991:243.

Golub ES, Green DR. *Immunology: a synthesis*. Boston, Massachusetts, USA: Sinauer Associates, 1991.

Grey HM, Sette A, Buus S. How T cells see antigen. In: Paul WE, ed. *Immunology: recognition and response*. New York: WH Freeman (Scientific American Inc), 1991:47–57.

Habeshaw J, Hounsell E, Dalgleish A. Does the HIV envelope induce a chronic graft-versus-host-like disease? *Immunology Today* 1992;**13**:207.

Irving BA, Weiss A. The cytoplasmic domain of the T cell receptor ζ chain is sufficient to couple to receptor-associated signal transduction pathways. *Cell* 1991;**64**:891.

June CH, Fletcher MC, Ledbetter JA, Schieven GL, Siegel JN, Philips AF, *et al.* Inhibition of tyrosine phosphorylation prevents T cell receptor-mediated signal transduction. *Proc Natl Acad Sci USA* 1990;**87**:7722.

Koyasu S, D'Adamio L, Clayton LK, Reinherz EL. T cell receptor isoforms and signal transduction. *Curr Opin Immunol* 1991;**3**:2.

Leder P. The genetics of antibody diversity. In: Paul WE, ed. *Immunology: recognition and response*. New York: WH Freeman (Scientific American Inc), 1991:20–34.

Marrack P, Kappler J. The T cell and its receptor. In: Paul WE, ed. *Immunology: recognition and response*. New York: WH Freeman (Scientific American), 1991:35–46.

Sheer D. Chromosomes and cancer. In: Franks LM, Teich NM, eds. *Introduction to the cellular and molecular biology of cancer*. Oxford: Oxford Medical Publications, 1991:269.

Samelson LE, Harford HB, Klausner RD. Identification of the components of the murine T cell antigen complex. *Cell* 1985;**43**:223.

Samelson LE, Philips AF, Luong ET, Klausner RD. Association of the *fyn* protein-tyrosine kinase with the T cell antigen receptor. *Proc Natl Acad Sci USA* 1990;**87**:4358.

Steinman L, Oksenberg JR, Bernard CCA. Association of susceptibility to multiple sclerosis with TCR genes. *Immunol Today* 1992;**13**:49.

Terhorst CB, Alarcon J, Spits H. T lymphocyte recognition and activation. In: Hames BD, Glover DM, eds. *Molecular immunology*. Oxford: IRL, 1989:145.

Veillette A, Davidson D. Src-related protein tyrosine kinases and T cell receptor signalling. *Trends Genet* 1992;**8**:61.

Williams AF, Barclay AN. The immunoglobulin superfamily-domains for cell surface recognition. *Annu Rev Immunol* 1988;**6**:381.

Monoclonal antibodies in medicine

Robert E Hawkins, Meirion B Llewelyn, Stephen J Russell

In 1975 George Köhler and César Milstein of the Medical Research Council Laboratory of Molecular Biology in Cambridge described an elegant system of obtaining pure antibodies of known specificities in large amounts (fig 1).[1] An astute reporter from the BBC World Service immediately recognised the potential of the discovery of monoclonal antibodies but it was some time before they were widely used. In the original method, a mouse is immunised repeatedly with the desired antigen and the spleen, which contains proliferating B cells, its removed. B cells normally die in culture, but can be immortalised by fusion with a non-secretory myeloma cell. The resulting hybridoma can then secrete large amounts of the antibody encoded by its B cell fusion partner. Supernatants from the hybrids that survive the selection procedure outlined in figure 1 are tested for binding to the original antigen.

Alan Williams showed in 1977 that monoclonal antibodies could be raised against biologically interesting molecules[2] and this triggered the development of a stream of useful monoclonal based diagnostic procedures. It soon became apparent, however, that rodent monoclonal antibodies were unsuitable or ineffective for treating humans because (*a*) they initiate human defence systems poorly (their fine structure differs significantly from that of human antibodies and in many cases the Fc portion is virtually unseen by human Fc receptors and complement proteins, resulting in failure to initiate human defence mechanisms (effector functions)) and (*b*) they are themselves the target of an immune response that can greatly shorten their circulating half life. One solution is to produce human monoclonal antibodies, but this is difficult for several reasons. Firstly, human B cells immortalised by fusion with

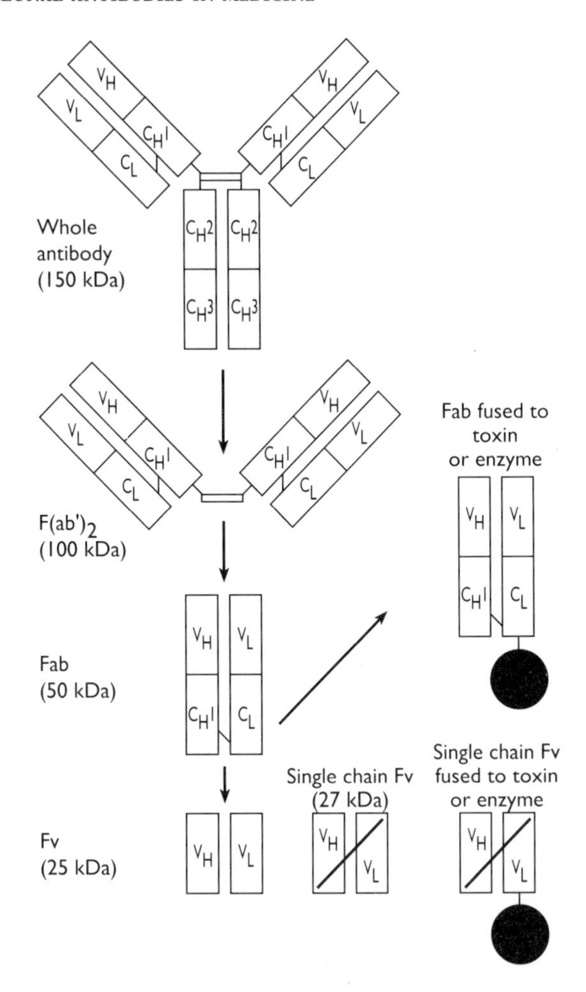

FIG 1—Antibody fragments. Limited proteolytic cleavage of IgG removes the Fc portion yielding bivalent $F(ab')_2$ (100 kDa) or monovalent Fab antigen binding fragments. Cloned T genes for Fab can be expressed in mammalian cells, yeast, or bacteria to produce functional recombinant Fab molecules. Toxins and enzymes can be fused to the recombinant Fab by genetic engineering. Smaller antibody fragments (Fv) can be produced in bacteria by coexpression of cloned V_H and V_L genes and stability is increased when V_H and V_L are linked (genetically) by a short peptide (single chain Fv or scFv). This can also be linked to toxins genetically

a myeloma cell or by infection with Epstein-Barr virus[3] are unstable and can rapidly lose their capacity for antibody production. Even when they are successfully transformed with Epstein-

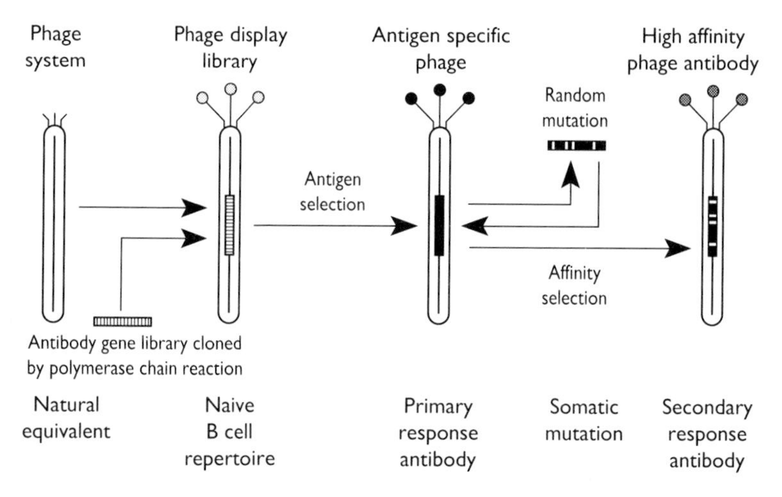

FIG 2—In vitro antibodies compared with the natural humoral immune system. The phage system allows all aspects of the humoral immune system to be mimicked in vitro

Barr virus antibody yields tend to be low and the virus usually transforms IgM secreting cells that produce lower affinity antibodies. Secondly, hyperimmunisation of human subjects is problematic, particularly against self antigens. In vitro immunisation—that is, antigenic stimulation of cultured human lymphocytes—provides a partial solution to this problem, but antibody production is still unstable after immortalisation. The third problem is that the most convenient source of human lymphocytes is the peripheral blood, a poor source of B cells producing specific high affinity antibodies. Such cells are more abundant in the spleen, bone marrow, or lymph nodes.

Fortunately, through developments in basic sciences, solutions to these problems are emerging and the promise of human monoclonal antibody therapy is at last becoming a reality. The immunogenicity of rodent monoclonal antibody can now be reduced, to improve their ability to recruit natural effector functions and increase their affinity. Many different antibody fragments can be produced and linked to various effector functions. More recently methods have begun to emerge for generating and screening large libraries of human antibodies entirely in vitro.

Chemical modification

Regardless of how a monoclonal antibody is produced, it may be desirable to tailor it to suit its intended application better. This can be achieved through chemical or genetic approaches. Figure 2 shows some of the possible modifications.

Proteolytic cleavage

Controlled proteolytic cleavage of a purified monoclonal antibody gives several smaller fragments that can be separated chromatographically. Cleavage was important in the early elucidation of antibody structure and structure-function relations but also has therapeutic implications. Antibody fragments are sometimes preferable to intact antibodies as they have a shorter circulating half life and may penetrate tissues more rapidly.

Chemical coupling

Monoclonal antibodies and antibody fragments can be conjugated chemically to a variety of substances, including plant and bacterial toxins, enzymes, radionuclides, and cytotoxic drugs. In this way, an ineffective rodent antibody or antibody fragment may be armed with a potent effector mechanism. Fragments coupled to radioactive elements can also be used for in vitro imaging or cancer therapy. However, the chemical coupling processes can be inefficient or give rise to unstable products, and repeated cycles of antibody purification, modification, and repurification are time consuming and costly.

Genetic modification

In contrast to a hybridoma or the protein it secretes, an antibody gene is highly versatile. It can be cut, joined to other genes, mutated randomly or non-randomly, and expressed in various cell types. The genes can be introduced into appropriate plasmid vectors and transfected into mammalian cells, bacteria, insect cells, or even plant cells for protein expression.

Antibody fragments

Antibody fragments similar to those generated by proteolytic cleavage can be generated from shortened versions of the heavy and light chain genes. The modified genes can then be transfected into bacteria[4] or mammalian cells where they produce functional

Fv (variable fragment) or Fab (antigen binding fragment) which is then easily purified. Bacteria are unsuitable hosts for producing complete monoclonal antibodies because the Fc domain of antibodies produced in this way is non-functional.

Antibody fusion proteins

Genetic engineering has been used to create chimeric molecules in which the variable domain of an antibody is genetically linked to an unrelated protein. In this way enzymes, toxins, and cytokines, for example, can be given novel binding specificities and can be produced in bacteria.[5] The approach is made easier when the antibody moiety is expressed as a single chain Fv (scFv) molecule in which the heavy and light chain V domains are linked by a short peptide that does not seriously affect antigen binding. The antibody can then be expressed as a single protein from a single gene rather than as two chains that must subsequently associate somewhat inside the cell. Several bacterial toxin-scFv fusion proteins have been produced, and because they are smaller than intact antibodies, it is hoped they will prove able to penetrate tumours more efficiently.

Antibodies can also be expressed on the surface of cytotoxic effector cells, redirecting them to kill novel targets. This approach has been used to redirect T cells[6] and has many potential applications to improve cellular immunotherapy.[7]

Humanisation

Chimeric antibodies—Genetic manipulation can also be used to make chimeric antibodies—that is, antibodies with rodent variable domains for antigen binding and human constant regions for recruiting effector functions.[8] The molecule is largely human but binds with the specificity of the parent monoclonal. Chimerisation enhances effector functions,[9] but a significant part of the molecule is still of rodent origin and recent human trials have shown that over half of humans mount an antimouse response after receiving a chimeric antibody.[10]

CDR grafted antibodies—Structural analysis of antibody-antigen complexes shows that the antigen binding surface of the antibody is formed by six hypervariable loops of amino acids called complementarity determining regions (CDRs). These loops are mounted on relatively constant framework regions and by genetic manipulation can be transplanted from a rodent antibody on to a human framework. This produces a CDR grafted or humanised antibody with the same specificity as the rodent monoclonal antibody from which the loops were grafted.[11] The process usually reduces the affinity of the antibody, but mutations can be made to restore full binding. Several antibodies have now been humanised[12] and one has already been used with clear therapeutic benefit.[13]

Bispecific antibodies

Bispecific antibodies have two antigen binding sites, each with a different binding specificity. Conventionally, they have been produced by fusing two hybridoma lines to make a hybrid hybridoma[14] or by chemical cross linking of antibody fragments. Because of random pairing of heavy and light chains and of heavy chain-light chain heterodimers, on average less than 10% of the IgG secreted by a cell expressing two antibody genes displays both of the required specificities. Genetic techniques allow the production of constructs which facilitate the association of non-identical species.[15] Bispecific monoclonal antibodies have been used to cross link cytotoxic effector cells to targets that they would not otherwise recognise—for example, tumour cells—and the approach has been used with apparent benefit in the treatment of malignant gliomas.[16] They can also be used to redirect toxins and enzymes to specific cellular targets.

Rapid cloning of antibody genes

No two antibody genes are identical so it might be expected that cloning each gene would be a tedious process. However, with the development of rapid methods based on the polymerase chain reaction (PCR) (see chapter on that) cloning functional rearranged V genes has become routine.

Although they differ in the middle, all antibody V genes are similar at either end, which allows construction of oligonucleotide sets whose sequences recognise and bind to the terminals of most V genes and prime the polymerase chain reaction.[17] Oligonucleotide sets are available for amplification of murine or human V genes.

It is now possible to amplify and rescue most of the antibody V genes from a diverse population of human or murine B cells, thereby generating an antibody gene library. This method works equally well with any type of B cell—resting B cells,[18] antibody secreting plasma cells, or memory B cells[19]—and the starting material may be either RNA or DNA.

Whatever the source of the antibody gene library, its usefulness depends on the availability of a convenient system for expressing the genes and selecting those that encode the best antibodies. Until recently, the best system available was suitable for screening no more than a million transfected colonies of *Escherichia coli*,[20] which is at least two orders of magnitude lower than the number of antibodies screened by an intact immune system. However, with the arrival of phage antibodies (see below),[21] libraries containing at least 10^8 different antibodies can now be screened.

Phage antibodies

Filamentous bacteriophages (hereafter referred to as phages) are pencil shaped viruses that infect bacteria. They attach to the surface of bacterial cells and inject their single stranded DNA genome through the cell wall. The infected bacterium does not die but continues to divide, distributing copies of the viral genome on to its progeny, which assemble and extrude perfect replicas of the invading phage. After overnight incubation, one millilitre of the bacterial culture supernatant contains over 10^{11} progeny phage particles.

At one tip of the phage are a few (probably three) copies of a protein (gene III protein). This protein mediates the initial attachment of the phage to a bacterial cell. To make a phage antibody the gene for an antibody fragment is fused precisely to one end of gene III on the phage genome. When this modified phage DNA is transfected into a bacterial cell, the cell produces and extrudes progeny phage particles that not only display the appropriate antibody at their tip (in fusion with the gene III protein) but also contain a single copy of the antibody gene and are still able to infect bacteria almost as efficiently as unmodified phages. Antibodies displayed on the surface of phages are fully functional and will still bind their antigen specifically. Phage antibodies with the desired specificity can be purified from a mixed population because of their ability to bind antigen.[21]

Thus a phage antibody is the functional in vitro equivalent of a resting B cell. It contains an antibody gene and displays the corresponding functional antibody on its surface. It can be selected for its ability to recognise a particular antigen, whereupon it can be amplified by growth in bacterial culture. It is then simple to rescue the antibody gene, which can be used to produce large amounts of soluble antibody (like the plasma cell) or simply stored in the freezer (acting similarly to the memory cell).

Phage antibody libraries

The intact humoral immune system is essentially a large library of antibodies. After challenge with antigen the most suitable antibodies are selected, amplified, and affinity matured. If the whole process could be reproduced entirely in vitro, production of high affinity human monoclonal antibodies might be greatly simplified.[22] With phage antibody libraries (see below) this goal is fast becoming a reality and these should increase the pace at which antibody therapy develops.

If a phage antibody can be likened to a B cell, then a phage antibody library is the in vitro equivalent of the humoral immune system (fig 3). A phage antibody library is constructed by ligating an antibody gene library (amplified by the polymerase chain reaction) into the appropriate site on purified phage DNA. The ligated DNA is transfected into bacteria, which then manufacture large numbers of phage antibodies, and those with the desired binding specificies are selected by using soluble tagged antigen or an antigen coated surface. Although still in its infancy, phage technology looks set to make a major impact on antibody development.

Future of antibody engineering

The technological trickery of antibody engineering is advancing more rapidly than it can be tested in therapeutic models. This presents a problem for pharmaceutical companies because in the time it takes them to scale up production methods for their most promising therapeutic monoclonal antibody, both the production method and the antibody may have been superseded. Notwithstanding, several companies have taken the plunge, and monoclonal antibodies are beginning to trickle into clinical practice. This trickle will probably soon become a flood and, faced with a plethora of cleverly conceived and constructed, but competing reagents, clinicians will benefit from an understanding of some of the principles of antibody therapy.

The 1990s will be a testing time for monoclonal antibodies. Potential clinical applications include the treatment of cancer, autoimmune disease, transplant rejection, viral infection, and toxic shock. The Centre for Exploitation of Science and Technology has estimated that the total world market for monoclonal antibodies will reach $1000 million by 1994, rising to $6000 million by the year 2000.[23] It remains to be seen whether the clinical promise of monoclonal antibodies will be realised on such a grand scale, but antibody therapy is likely to be much in evidence in many clinical settings over the next few years. Clinicians will therefore need to familiarise themselves with some of the issues relating to use of clinical antibodies.

Target antigens

Antibodies can neutralise toxins; block the interaction of growth factors, hormones, intercellular adhesion molecules, or viruses with their cognate cellular receptors; and coat bacteria, viruses, or cells, marking them for phagocytosis, antibody dependent cellular cytotoxicity, or complement mediated lysis.

Target antigens can therefore be circulating or on the cell surface. Selecting a suitable target for a given disease depends not only on the aims of treatment but on the precise tissue distribution of the target antigen, its function, and its fate after it has complexed with the therapeutic monoclonal antibody. Finding a suitable target antigen is probably the most important factor determining the ultimate success or failure of antibody therapy. Provided that the target has been well chosen it may be possible to modify the corresponding monoclonal antibody in various ways to enhance its therapeutic potential.

Pharmacokinetics

Infused antibodies are diluted almost immediately in the total plasma volume and then diffuse more slowly across the walls of small blood vessels into the interstitial fluid (distribution phase). The half life of the circulating antibody is determined by the rate at which it is metabolised and excreted (elimination phase). The degree to which the target antigen is bound varies with the total time of exposure, the concentration, and the kinetic properties of the monoclonal antibody.

Reaching the target

How easily infused monoclonal antibody can reach it depends on the target's location. Intravascular targets are readily accessible, but other targets are less easily reached because exit from the vascular system is restricted. To gain access to extravascular targets (for example, cancer cell surface antigens) the antibody must pass through the endothelial lining of a capillary or post-capillary venule. Smaller antibody fragments, particularly Fv reagents, penetrate the interstitial fluid space more readily than whole IgG. High molecular weight proteins can escape from the microvessels through gaps between adjacent endothelial cells, which are particularly abundant in inflamed tissues. The discontinuous endothelial lining of the sinusoidal circulations of liver, spleen, and bone marrow also allows free passage of macromolecules such as IgG.

Binding reaction

The degree to which a therapeutic antibody binds to its target antigen is governed by the concentration of antigen, the concentration of antibody to which it is exposed, the duration of exposure, and the intrinsic properties of antibody and antigen that determine their rates of association and dissociation.

The equation is simple when attempting, for example, to neutralise a circulating toxin. For cell surface antigens the analysis is less simple. The concentration of antibody to which the target cells are exposed and the duration of exposure are determined by the rate at which the antibody enters the interstitial fluid and the rate at which it is eliminated from the body. Also, cell associated targets are effectively multivalent so that affinity is no longer the only factor determining the rate of dissociation of cell bound antibody. Provided that the target antigen is expressed at sufficiently high density, bivalent molecules such as intact IgG or $F(ab)_2$ can bind with much greater avidity than can the smaller univalent Fab and Fv antibody fragments. Complexed antibody may also be taken into the cytoplasm of the target cell, effectively preventing further dissociation. Additionally, a cluster of many cells displaying the same target antigen (a tumour deposit, for example) may behave as an antigen "sink" from which the antibody escapes only very slowly. This is because dissociation of a bound antibody molecule will be followed immediately by rebinding to the same or a neighbouring cell.

Antibody clearance

Ultimately, all infused antibody will be eliminated from the body. The Fc portion of IgG is thought to determine its catabolic rate, which (in humans) is faster for murine antibodies than for human antibodies.[25] Smaller antibody fragments pass relatively easily from the glomerular capillaries into the renal tubules and are rapidly excreted unchanged in the urine; this greatly shortens their plasma half life. Thus the circulating half life of IgG is measured in days, and that of single chain Fv fragments (scFv) in minutes, while $F(ab)_2$ and Fab fragments have intermediate half lives.[24]

Clearance of antibody that has been retained in the tissues is slower. Retention of antibodies by tissue because of specific interaction of the antibody with its target antigen is welcome, but non-specific binding to homologous or non-homologous antigens also occurs. Moreover, IgG may be retained in liver, spleen, and bone marrow through the interaction of its Fc portion with Fc receptors on resident macrophages. Fab and $F(ab)_2$ fragments tend to accumulate in the kidneys.[3] Persistence of antibodies in normal host tissues may be troublesome, leading, for example, to excessive toxicity of a radiolabelled therapeutic cancer antibody.

Effector mechanisms

When the goal of treatment is neutralisation of a toxin or blockade of a ligand-receptor interaction the therapeutic antibody requires no special effector domain and, depending on the effective valency of the target antigen, should function well as a monovalent (single chain Fv or Fab) or bivalent ($F(ab)_2$) fragment. More commonly, however, the aim is to destroy a specific population of target cells. Phagocytosis, antibody dependent cellular cytotoxicity, and complement fixation are the natural effector pathways activated by the Fc portion of cell bound antibody. Smaller antibody fragments without an Fc portion (single chain Fv, Fab, $F(ab)_2$) can be artificially given alternative effector mechanisms including radioactive metals, plant and bacterial toxins, enzymes, and cytotoxic drugs.

The most suitable effector mechanism depends on several factors. A high density of IgG on the target cell is required to activate complement because it is initiated by cross linking the Fc portion of two adjacent cell bound antibody molecules. Moreover, Fc mediated recruitment of phagocytes, antibody dependent cellular cytoxicity, and complement is not possible unless bound antibody stays on the surface of the target cells.

For some conditions it may be more appropriate to use antibodies with artificially linked effector functions. Cells that rapidly

Effector functions for antibody targeted therapy

Blocking	Ligand-receptor interactions
	• Cell adhesion
	• Virus attachment
	• Cytokine stimulation
Natural	Fc mediated
	• Complement fixation
	• Antibody dependent cellular cytotoxicity
	• Phagocytosis
Artificial	Toxins
	• Plant toxins
	• Bacterial toxins
	Radioisotopes
	Cytotoxic drugs
	Enzymes
	• Prodrug activation
	• Direct toxicity
Bifunctional	Cross linking
	• Cytotoxic effectors to targets
	• Enzymes or toxins to targets

internalise bound antibody or express the target antigen at low density may be killed more effectively by antibodies conjugated to drugs, toxins, or radionuclides. Radioimmunoconjugates also have the advantage that their radiation can penetrate several cell diameters into the tissues—this may be useful for cancer therapy as the monoclonal antibody cannot penetrate deep into the tumour. The box summarises the available antibody effector mechanisms.

Clinical use of monoclonal antibodies

Immunosuppression

Monoclonal antibodies offer a realistic alternative to immunosuppressive drugs, and this is perhaps their most useful current application. Potential targets for immunosuppressive monoclonal antibodies include lymphocyte differentiation antigens, cytokines, cytokine receptors, and cell adhesion molecules.

The first monoclonal antibody to be approved for human therapy (OKT3) is an immunosuppressive murine reagent which binds to T lymphocytes and is useful for treating rejection of renal transplants.[26,27] In common with many other immunosuppressive antilymphocyte monoclonal antibodies it does not stimulate a strong antimouse response. The toxicity of OKT3 is worse with the first dose, which triggers release of cytokines from targeted cells and leads in some cases to hypotension, weight gain, and breathlessness, progressing occasionally to pulmonary oedema. Many other immunosuppressive monoclonal antibodies have been shown to have activity in humans. Among the most promising are antibodies against the lymphocyte antigens CD4, Tac, and CDw52 (see below), all of which have now been humanised by CDR grafting,[28–30] and several monoclonal antibodies that block adhesion of immune and inflammatory cells.

Monoclonal antibodies against CD4 inhibit the function of helper T cells and have been used with varying success to treat acute rejection of renal allografts, rheumatoid arthritis, inflammatory bowel disease, systemic lupus erythematosus, psoriasis, relapsing polychondritis, systemic vasculitis, and mycosis fungoides.[31–37] Tac monoclonal antibodies recognise high affinity interleukin 2 receptors of activated lymphocytes and do not bind to resting lymphocytes. They can therefore block ongoing antigen specific immune responses highly specifically without damaging resting lymphocytes. Murine Tac monoclonal antibodies were shown to prevent early rejection of renal allografts, but antimouse responses were detected in 81% of patients after one month of treatment.[38] Humanised Tac antibody (Tac-H) was recently compared with the murine antibody in primates given cardiac allografts.[39] The humanised antibody had a longer circulating half life (103 *v* 38 h), was less immunogenic (0% *v* 100% antiantibody responses before day 33), and produced a longer graft survival than the murine antibody.

Immunosuppressive monoclonal antibodies will undoubtedly contribute to the therapeutic options against autoimmune disease and rejection of transplants but their precise role has yet to be defined. More detailed understanding of the underlying immunopathogenic mechanisms in many autoimmune conditions and vasculitic states will help future exploration of the therapeutic potential of monoclonal antibodies.

Infection

Agammaglobulinaemic patients suffer from recurrent bacterial sinopulmonary infection, meningitis, and bacteraemia.[40] Viral infections are no more severe in patients with agammaglobulinaemia than in healthy people, suggesting that T cells are the most important initial defence, but lasting immunity is lacking so multiple bouts of chickenpox and measles may occur. These observations suggest that antibodies should be able to prevent bacterial and viral infections. Indeed, regular administration of purified pooled human immunoglobulin provides good protection for patients with agammaglobulinaemia (and hypogammaglobulinaemia or dysgammaglobulinaemia).[41]

Polyclonal human immunoglobulin preparations have been used for many years to treat and prevent several viral diseases including hepatitis A and B, chickenpox, measles, and cytomegalovirus infection. Several antiviral and antibacterial monoclonal antibodies are now under development for human risks. For example, humanised versions of monoclonal antibodies to herpes simplex virus[42] and respiratory syncytial virus[43] have been prepared and human antibodies to HIV have been isolated by screening phage libraries.[44] Antiviral monoclonal antibodies can block attachment and penetration of viruses, opsonise virus and virus infected cells for phagocytosis or antibody dependent cytoxicity, and mediate complement lysis of enveloped virus particles or infected cells. Cocktails of monoclonal antibodies will probably give greater benefit than single reagents.

However, as T cells and not antibodies seem to be essential for eradicating established viral infections, it can be argued that antibodies are unlikely to be useful in treating these conditions. Moreover, there is evidence that certain viral infections may be enhanced by antiviral antibodies, which can facilitate Fc receptor mediated viral entry into macrophages and some other cells.

Toxic states

The use of monoclonal antibodies to treat septic shock has been reviewed recently.[45] Endotoxin, a lipopolysaccharide component of the bacterial cell wall, damages vascular endothelium, which triggers a cascade of events that leads to septic shock. Because the target antigen is intravascular, IgM monoclonal antibodies can be used. HA-1A, a human IgM anti-endotoxin monoclonal antibody, reduced 28 day mortality by 39% in 105 patients with Gram negative bacteraemia.[46] Although the result seems impressive, the antibody is expensive and it is difficult to design protocols that avoid treating large numbers of patients who subsequently prove not to have had Gram negative bacteraemia. Moreover, the initial study has been seriously criticised and a second placebo controlled clinical trial of HA-1A has been recommended to assess whether the antibody should be widely used.[47]

Tumour necrosis factor is one of the central mediators of septic shock, and monoclonal antibodies against it are protective in animal models. A phase I clinical trial of one monoclonal antibody against tumour necrosis factor confirmed its safety, but its efficacy has yet to be shown in humans.[48] Besides the obvious examples of tetanus and diphtheria, other toxic states that may be amenable to monoclonal antibody therapy include drug overdosage, chemical poisoning, and snake or spider bites. Already digoxin Fab fragments are well established for the management of digoxin overdose and monoclonal antibodies are being developed for neutralising tricyclic antidepressants.[49]

Cancer (solid tumours)

Monoclonal antibodies against cancers have been used for both imaging and treatment. There are numerous possible target antigens, which fall into several broad categories (box). Except in a few cases, unique tumour specific antigens have not been identified, and studies have focused on target antigens that are present to a greater or lesser degree on some normal host tissues. Examples include oncofetal antigens such as carcinoembryonic antigen and α fetoprotein, epidermal growth factor receptors, carbohydrate antigens, and components of the extracellular matrix such as mucin. Radioimmunoconjugates accumulate in tumour deposits well enough to produce reasonable images,[50] although the image is not yet good enough seriously to challenge conventional imaging methods such as computed tomography. Treatment of cancer with monoclonal antibodies has so far been disappointing.[51] Early studies used immunogenic murine monoclonal antibodies that could not recruit human effector functions. Humanising these monoclonal antibodies or linking them to radioisotopes, toxins, and drugs (which may increase immunogenicity) has so far had little impact on their therapeutic efficacy and can produce serious toxicities. However, it would be inappropriate to discount the potential of these alternative killing mechanisms.

An important limiting factor is the inability of infused monoclonal antibodies to reach the target cells. Monoclonal antibodies have good access to the tumour surface as the surface blood vessels of a tumour deposit are relatively leaky to macromolecules but the branches of these vessels, which penetrate the tumour parenchyma, are not.[52,53] Once on the surface, however, they meet an impenetrable wall of tumour cells held together by tight intercellular junctions, which makes access to deeper parenchymal regions

Target antigens for monoclonal antibodies against cancer

Unique to tumour	Immunoglobulins T cell receptors Mutated cell surface proteins
Relative abundance in tumour	Growth factor receptors Oncofetal antigens Dead cell markers
Confined to tumour and non-essential normal tissues	Differentiation antigens
Stromal targets	Endothelial activation markers Fibroblast activation markers

of the tumour poor. It was hoped that smaller versions of the antibody molecule—for example, Fv_1, single chain Fv—would escape more readily from penetrating vessels and permeate better through the parenchymal regions of the tumour. However, the early signs are that their lack of avidity (they are univalent) and rapid renal excretion result in lower absolute tumour uptake despite a better tumour to normal tissue ratio.[25]

The toxicity of cancer monoclonal antibodies has been variable. With unmodified murine monoclonal antibodies fever, rigors, nausea, and vomiting are common after the initial doses, immediate hypersensitivity reactions can occur, and symptoms secondary to circulating immune complexes are sometimes seen after prolonged treatment. Radioimmunoconjugates usually cause appreciable toxicity to normal bone marrow, and immunotoxins can cause the vascular leak syndrome.

Antibody dependent enzyme prodrug therapy (ADEPT) is a promising research prospect.[54] With this technique an antibody-enzyme conjugate is administered, which localises to tumour deposits. After a few days, during which non-specifically bound monoclonal antibody is cleared, an inactive prodrug is administered. The prodrug is converted by monoclonal antibody-linked enzyme in tumour deposits to an active tumoricidal drug that is small enough to permeate the deeper regions of the tumour. Human and humanised antibodies with improved affinity and specificity are likely to be used increasingly in the future. One recent animal experiment has shown that improved affinity can give improved anticancer activity.[55] Phage technology could help develop appropriate antibodies.[56] Cocktails of monoclonal antibodies may give better results than single antibodies (see below).

Haematological malignancies

Monoclonal antibody treatment for haematological malignancies has been more successful than that for solid tumours. Activity against disease in bone marrow and spleen has been notable, with nodal disease responding less readily.[25] One possible explanation is that the sinusoidal circulations of responsive organs are easily permeated by immunoglobulins. Also the bone marrow and spleen are rich in host effector cells that can recognise and kill targets coated by monoclonal antibodies.

Murine anti-idiotypic monoclonal antibodies have been raised against unique surface immunoglobulins and T cell receptors expressed respectively on B and T cell malignancies. The results of treatment are encouraging[56,57] but monoclonal antibodies have to be tailor made for each patient. Alternative targets include a wide array of leucocyte differentiation antigens. B cell malignancies, for example, have been treated with monoclonal antibodies against the lymphocyte antigens CD19, CD22, CD37, and CDw52.[25,58,59] The monoclonal antibodies inevitably destroy some normal lymphocytes, but these are regenerated from stem cells, which are not attacked. Transient antibody related immunosuppression can, however, be troublesome.

Natural effector functions are effective against several haematological malignancies. CAMPATH-1G is a rat IgG2b monoclonal antibody that recruits human complement and antibody dependent cytotoxicity and binds to an antigen (CDw52) present on most normal and malignant lymphocytes. Of 29 patients with lymphoid malignancies who received CAMPATH-1G, nine attained complete remissions, although disease in lymph nodes was generally resistant to treatment.[25] A CDR grafted version of this antibody (CAMPATH-1H) was the first humanised monoclonal antibody to enter clinical trials and induced complete remissions in two patients with B cell non-Hodgkin's lymphoma, one with lymph node disease.[60] More extensive clinical testing of CAMPATH-1H is currently under way.

Immunotoxins and radioimmunoconjugates have shown activity against lymphoma, but direct comparisons of alternative effector mechanisms on a single monoclonal antibody have not yet been made. Polyclonal antiferritin antisera have been shown to target Hodgkin's disease deposits more efficiently than antiferritin monoclonal antibodies,[61] which suggests again that monoclonal antibody cocktails may be the best way forward.

Other applications

Monoclonal antibodies are being developed for imaging of infarcted myocardium (antimyosin), deep venous or arterial thromboses (antifibrin), and foci of infection of inflammation. Antirhesus monoclonal antibodies have been made for treating rhesus haemolytic disease and antiplatelet monoclonal antibodies for prevention of intravascular thrombosis. Monoclonal antibody enzyme conjugates targeted at blood clots are also under development as novel fibrinolytic reagents.

Conclusions

We have progressed considerably since the early days of monoclonal antibody therapy but there is still much to learn. Human (or humanised) monoclonal antibodies are preferable to rodent monoclonal antibodies for most applications. Cocktails of monoclonal antibodies should be more effective than single antibodies, and production of such cocktails will be helped by the advent of human phage antibody libraries. Enhancement of an antibody's affinity is now possible by phage technology,[35] and the early signs suggest that it should improve therapeutic efficacy. Definitive studies comparing the clinical efficacy of various natural and artificial effector functions are needed, and there is scope for boosting natural effector mechanisms with lymphokine therapy. For the future, antibodies and antibody genes may be used increasingly to redirect cytotoxic cells or for targeted delivery of genes and other drugs wrapped up in viruses or liposomes.

1 Kohler G, Milstein C. Continuous cultures of fused cells secreting antibody of predefined specificity. *Nature* 1975;**256**:495–7.

2 Williams AF, Galfre G, Milstein C. Analysis of cell surfaces by xenogenic myeloma-hybrid antibodies: Differentiation antigens of rat lymphocytes. *Cell* 1977;**12**:663–73.

3 Steinitz M, Izak G, Cohen S, Ehrenfeld M, Flechner I. Continuous production of monoclonal rheumatoid factor by EBV-transformed lymphocytes. *Nature* 1980;**287**:443–5.

4 Skerra A, Pluckthün A. Assembly of a functional immunoglobulin Fv fragment in Escherichia coli. *Science* 1988;**240**:1038–41.

5 Gross G, Waks T, Eshhar Z. Expression of immunoglobulin-T-cell receptor chimeric molecules as functional receptors with antibody-type specificity. *Proc Natl Acad Sci USA* 1989;**86**:10024–8.

6 Rosenburg SA. The immunotherapy and gene therapy of cancer. *J Clin Oncol* 1992;**10**:180–99.

7 Milstein C, Cuello AC. Hybrid hybridomas and their use in immunohistochemistry. *Nature* 1983;**305**:537–40.

8 Neuberger MS. Williams GT, Mitchell EB, Jouhal SS, Flanagan JG, Rabbitts TH. A hapten-specific chimaeric IgE with human physiological effector function. *Nature* 1985;**314**:268–70.

9 Brüggemann M, Williams GT, Bindon CI, Clark MR, Walker MR, Jefferis R, *et al.* Comparison of the effector functions of human immunoglobulins using a matched set of chimeric antibodies. *J Exp Med* 1987;**166**:1351–61.

10 Meredith RF, Khazaeli MB, Plot WE, Saleh MN, Liu T, Allen LF, *et al.* Phase I trial of iodine-131-chimeric B72.3 (human IgG4) in metastatic colorectal cancer. *J Nucl Med* 1992;**33**:23–9.

11 Jones PT, Dear PH, Foote J, Neuberger MS, Winter G. Replacing the complementarity-determining regions of a human antibody with those from a mouse. *Nature* 1986;**321**:522–5.

12 Russell SJ, Llewelyn MB, Hawkins RE. The human antibody library: entering the next phage. *BMJ* 1992;**304**:585–6.

13 Hale G, Clark MR, Marcus R, Winter G, Dyer MJS, Phillips JM, *et al.* Remission induction in non-Hodgkin lymphoma with reshaped monoclonal antibody CAMPATH-1H. *Lancet* 1988;**ii**:1394–9.

14 Pastan I, Fitzgerald D. Recombinant toxins for cancer treatment. *Science* 1991;**254**:1173–7.

15 Kostelny SA, Cole MS, Tso JY. Formation of a bispecific antibody by use of leucine zippers. *J Immunol* 1992;**148**:1547–53.

16 Nitta T, Sato K, Yagita H, Okumura K, Ishi S. Preliminary trial of specific targeting therapy against malignant glioma. *Lancet* 1990; **335**:368–71.

17 Orlandi R, Güssow DH, Jones PT, Winter G. Cloning immunoglobulin variable domains for expression by the polymerase chain reaction. *Proc Natl Acad Sci USA* 1989;**86**:3833–7.

18 Marks JD, Hoogenboom HR, Bonnett TP, MacCafferty J, Griffiths AD, Winter G. By-passing immunization: human antibodies from V-gene libraries displayed on bacteriophage. *J Mol Biol* 1991;**222**:581–97.

19 Hawkins RE, Winter G. Cell selection strategies for making antibodies from variable gene libraries: trapping the memory pool. *Eur J Immunol* 1992;**22**:867–70.

20 Huse WD, Sastry L, Iverson S, Kang AS, Alting-Mees M, Burton DR, *et al.* Generation of a large combinatorial library of the immunoglobulin library in phage lambda. *Science* 1989;**246**:1275–81.

21 MacCafferty J, Griffiths AD, Winter G, Chiswell DJ. Phage antibodies: filamentous phage displaying antibody variable domains. *Nature* 1990;**348**:552–4.

22 Winter G, Milstein C. Man-made antibodies. *Nature* 1991;**349**:293–9.

23 Savin J. *The value of antibody engineering technology to the UK*. London: Centre for Exploitation of Science and Technology, 1990.

24 Milenic DE, Yokota T, Filpula DR, Finkelman MAJ, Dodd SW, Wood JF, *et al.* Construction, binding properties, metabolism, and tumor targeting of a single-chain Fv derived from the pancarcinoma monoclonal antibody CC49. *Cancer Res* 1991;**51**:6363–71.

25 Waldmann TA. Monoclonal antibodies in diagnosis and therapy. *Science* 1991;**252**:1657–62.

26 Ortho Multicentre Transplant Study Group. A randomised trial of OKT3 monoclonal antibody for acute rejection of cadaveric renal transplants. *N Engl J Med* 1985;**13**:337–42.

27 Carpenter CB. Immunosuppression in organ transplantation. *N Engl J Med* 1990;**322**:1224–6.

28 Gorman SD, Clark MR, Routledge EG, Cobbold SP, Waldmann H. Reshaping a therapeutic CD4 antibody. *Proc Natl Acad Sci USA* 1991;**88**:4181–5.

29 Queen C, Schneider WP, Selick HE, Payne PW, Landolfi NF, Duncan JF, *et al.* A humanized antibody that binds to the interleukin 2 receptor. *Proc Natl Acad Sci USA* 1989;**86**:10029–33.
30 Riechmann L, Clark M, Waldmann H, Winter G. Reshaping human antibodies for therapy. *Nature* 1988;**332**:323–7.
31 Reinke P, Miller H, Fietze E, Herberger D, Volk HD, Neuhaus K, *et al.* Anti-CD4 therapy of acute rejection in long-term renal allograft recipients. *Lancet* 1991;**338**:702–3.
32 Horneff G, Burmester GR, Emmrich F, Kalden JR. Treatment of rheumatoid arthritis with an anti-CD4 monoclonal antibody. *Arthritis Rheum* 1991;**34**:129–40.
33 Emmrich J, Seyfarth M, Fleig WE, Emmrich F. Treatment of inflammatory bowel disease with anti-CD4 monoclonal antibody. *Lancet* 1991;**338**:570–1.
34 Hiepe F, Volk HD, Apostoloff E, Baehr RV, Emmrich F. Treatment of severe systemic lupus erythematosus with anti-CD4 monoclonal antibody. *Lancet* 1991;**338**:1529–30.
35 Nicolas JF, Chamchick N, Thivolet J, Wijdenes NB, Morel P, Revillard JP. CD4 antibody treatment to severe psoriasis. *Lancet* 1991;**338**:321.
36 Van der Lubbe PA, Miltenburg AM, Breedveld FC. Anti-CD4 monoclonal antibody for relapsing polychondritis. *Lancet* 1991;**337**:1349.
37 Mathieson PW, Cobbold SP, Hale G, Clark MR, Oliveira DBG, Lockwood CM, *et al.* Monoclonal antibody therapy in systemic vasculitis. *N Engl J Med* 1990;**323**:250–4.
38 Sollilou J, Cantarovich D, LeMauff B, Giral M, Robillard N, Hourmant M, *et al.* Randomised controlled trial of monoclonal antibody against the interleukin 2 receptor (33B3.1) as compared with rabbit antithymocyte globulin for prophylaxis against rejection of renal allografts. *N Engl J Med* 1990;**322**: 1175–82.
39 Brown PS, Parenteau GL, Dirbas FM, Garsia RJ, Goldman CK, Bukowski MA, *et al.* Anti-Tac-H, a humanized antibody to the interleukin 2 receptor, prolongs primate cardiac allograft survival. *Proc Natl Acad Sci USA* 1991;**88**:2663–7.
40 Spickett GP, Misbah SA, Chapel HM. Primary antibody deficiency in adults. *Lancet* 1991;**337**:281–4.
41 Webster ADB. Intravenous immunoglobulins. *BMJ* 1991;**303**:375–6.
42 Co MS, Deschamps M, Whitley RJ, Queen C. Humanized antibodies for antiviral therapy. *Proc Natl Acad Sci USA* 1991;**88**:2869–73.
43 Tempest PR, Bremmer P, Lambert M, Taylor G, Furze JM, Karr FJ, *et al.* Reshaping a human monoclonal antibody to inhibit human respiratory syncytial virus infection in vivo. *Biological Technology* 1991;**9**:266–71.
44 Burton DR, Barbas III CF, Persson MAA, Koenig S, Chanock RM, Lerner RA. A large array of human monoclonal antibodies to type I human immunodeficiency virus from combinatorial libraries of asymptomatic seropositive individuals. *Proc Natl Acad Sci USA* 1991;**88**:10134–7.
45 Hinds CJ. Monoclonal antibodies in sepsis and septic shock. *BMJ* 1992;**304**:132–3.
46 Ziegler EJ, Fisher CJ Jr, Sprung CL, Straube RC, Sadoff JC, Foulke GE, *et al.* Treatment of Gram-negative bacteremia and septic shock with HA-1A human monoclonal antibody against endotoxin: a randomized, double blind, placebo-controlled trial. *N Engl J Med* 1991;**324**:429–36.
47 Warren HS, Danner RL, Munford RS. Sounding board. Anti-endotoxin monoclonal antibodies. *N Engl J Med* 1992;**326**:1153–7.
48 Exley AR, Cohen J, Buurman W, Owen R, Hanson G, Lumley J, *et al.* Monoclonal antibody to TNF in septic shock. *Lancet* 1990;**335**:1275–6.
49 Kulig K. Initial management of ingestions of toxic substances. *N Engl J Med* 1992;**326**:1677–81.

50 Order SE. Presidential address: systemic radiotherapy—the new frontier. *Int J Radiat Oncol Biol Phys* 1990;**18**:981–92.
51 Dvorak HF, Nagy JA, Dvorak AM. Structure of solid tumors and their vasculature: implications for therapy with monoclonal antibodies. *Cancer Cells* 1991;**3**:77–85.
52 Dvorak HF, Nagy JA, Dvorak JT, Dvorak AMI. Identification and characterization of the blood vessels of solid tumors that are leaky to circulating macromolecules. *Am J Pathol* 1988;**133**:95–109.
53 Bagshawe KD. Towards generating cytotoxic agents at cancer sites. *Br J Cancer* 1989;**60**:275–81.
54 Schlom J, Eggensperger D, Colcher D, Molinolo A, Houchens D, Miller LS, *et al.* Therapeutic advantage of high-affinity anticarcinoma radioimmunoconjugates. *Cancer Res* 1992;**52**:1067–72.
55 Hawkins RE, Russell SJ, Winter G. Selection of phage antibodies by binding affinity: mimicking affinity maturation. *J Mol Biol* 1992;**226**:889–96.
56 Brown SL, Miller RA, Horning SJ, Czerwinski D, Hart SM, McElderry R, *et al.* Treatment of B cell lymphomas with anti-idiotype antibodies alone and in combination with alpha interferon. *Blood* 1989;**73**:651–61.
57 Janson CH, Tehrani MJ, Mellstedt H, Wigzell H. Anti-idiotypic monoclonal antibody to a T cell chronic lymphatic leukaemia. *Cancer Immunol Immunother* 1989;**28**:225–32.
58 Vitetta ES, Stone M, Amlot P, Fay J, May R, Til M, *et al.* Phase I immunotoxin trial in patients with B-cell lymphoma. *Cancer Res* 1991;**51**: 4052–8.
59 Press OW, Eary JF, Badger CC, Martin PJ, Appelbaum FR, Levy R, *et al.* Treatment of refractory non-Hodgkin's lymphoma with radiolabelled MB-1 (anti-CD37) antibody. *J Clin Oncol* 1989;**7**:1027–38.
60 Hale G, Dyer MJS, Clark MR, Phillips JM, Marcus R, Riechmann L, *et al.* Remission induction in non-Hodgkin's lymphoma with reshaped human monoclonal antibody CAMPATH-1H *Lancet* 1988;**ii**:1394–9.
61 Vriesendorp HM, Herpst JM, Germack MA, Klein JLK, Leichner PK, Loudenslager DM, *et al.* Phase I-II studies of yttrium-labelled antiferritin treatment for end-stage Hodgkin's disease, including radiation therapy oncology group 87-01. *J Clin Oncol* 1991;**9**:918–28.

An introduction to cells

L Wolpert

Cells are the triumph of evolution.[1] The rest of evolution can be thought of as an elaboration on this masterpiece. In some ways they are more complex than the organs to which they give rise, with the possible exception of the brain, in that their behaviour reflects the integrated activity of about 50 000 genes, their products, and the complex biochemical and structural networks that result.

In this biochemical network there are two different time scales. The first, which responds to changes in seconds or fractions of a second, is that concerned with metabolism. The enzymes in the cell cytoplasm, including the mitochondria, catalyse molecules along narrowly defined reaction pathways such as the Krebs cycle, the synthesis of purines, or the breakdown of carbohydrates. Many of these reactions require or generate energy in the form of ATP. The speed of response of these metabolic pathways can be contrasted with those in the second system, which entails the synthesis of macromolecules such as nucleic acids and proteins. Here the response times are minutes to hours. Understanding the integration of these two interdependent pathways is a major problem in cell biology.

Proteins characterise cells

The character of a cell is determined by the proteins it contains. For example, the special feature of parenchymal liver cells is that they synthesise albumin, whereas lymphocytes synthesise globulin. The enzymes in a cell determine its metabolic pathways. The control of protein synthesis is thus central to the life of the cells and is controlled by nuclear cytoplasmic interactions. It should be noted that cell biochemistry is based on protein complexes, rather than single proteins acting individually.

The genes on the chromosomes in the cell nucleus dominate the life of the cell by controlling protein synthesis. But they are curiously passive. The DNA of the genes codes only for proteins or for RNA itself. Typically, a DNA sequence is transcribed into a messenger RNA (mRNA) sequence, which is then processed within the nucleus, non-coding sequences known as introns being removed. It is this processed RNA that enters the cytoplasm as mRNA, where it is translated into the amino acid sequence of the protein on ribosomes. Some of the RNA transcribed within the nucleus never reaches the cytoplasm, but the control in the nucleus is largely at the level of transcription itself. The genetic control of cell activity is entirely the result of the control of protein synthesis, and thus the genes do not exert an immediate effect on the life of the cell. If the nucleus is removed from the cell the metabolic pathways are unaffected until there is a change in enzyme concentration. This will depend on the stability of both the proteins and their mRNA.

DNA and function and form

As there is good evidence that the DNA is the same in almost all cells in the body of an individual, there must be mechanisms that turn genes on and off in specific cells. Embryonic development can thus be viewed in terms of the control of differential gene activities, DNA providing a genetic programme. As development proceeds different cell types with different protein compositions emerge. Not only must genes be turned on and off but this activity must also be stable so that it can be inherited when the cells multiply. Liver cells divide to give more liver cells.

Changes in DNA can alter cell behaviour. Mutations alter the sequence of the bases. This is what happens in sickle cell anaemia, in which the haemoglobin molecules have abnormal properties. A change in one of the codons of the gene for haemoglobin alters one amino acid, glutamine, to valine. This in turn changes the way the molecule folds and results in the haemoglobin molecules sticking together and forming rod-like structures, which deform the red cell, giving it its sickle shape. This alters the flexibility of the cell and so affects its passage through fine capillaries, which results in anaemia. Thus from a change in just one base there is a cascade of events leading to a physiologically abnormal condition.

Sickle cell anaemia is an excellent example of the relation between DNA and function and form. In contrast to the intimate relation between DNA and protein, the pathways whereby a protein affects the working of the cell and the body as a whole can be intricate, tortuous, and multifarious. In general these pathways are not known, and this is a major problem in cell biology. Even if we knew the complete sequence of the DNA of human cells we could not, at present, interpret it properly.

Role of cytoplasm

A central problem is in assessing what controls the transcription of particular genes. This involves protein transcription factors which are synthesised in the cytoplasm and then move to the nucleus. Such nuclear factors act on the promoters adjacent to particular structural genes and allow them to be transcribed in specific tissues. It is possible to attach the growth hormone gene (normally active only in the pituitary) to the elastase promoter (elastase is made only in the pancreas). When this construction is incorporated into the genome by injecting the DNA into the nucleus of a fertilised mouse egg then growth hormone is also made in the pancreas of the transgenic mouse.[2]

For some cells, at least, the pattern of gene expression is reversible.[3] Various different human cells, like liver cells, can be fused with mouse muscle cells. When the liver nucleus is exposed to muscle cytoplasm then genes specific for muscle begin to be transcribed and human muscle proteins synthesised. It seems that the pattern of gene expression remains sensitive to changes in nuclear factors.

The different proteins must be transported to the specific cellular compartments where they will be used: enzymes of the Krebs cycle to the mitochondria and degradative enzymes to liposomes. This is achieved by the extensive membrane bound system in the cytoplasm, particularly the endoplasmic reticulum and the Golgi apparatus. To achieve this sorting, specific amino acid sequences serve as address markers. For example, if the amino acid terminal end carries the so called signal sequence then the proteins are transferred across the membrane of the endoplasmic reticulum. Those that do not have the sequence remain in the cytosol. Further sorting then occurs so that, for example, some of those that enter the endoplasmic reticulum go through the Golgi apparatus to the plasma membrane or special secretory vesicles. It is clear that there is considerable organisation of the internal membrane system in the cytoplasm.

A major change in our concept of the cell relates to the structure of the cytoplasm or, more fashionably, the cytoskeleton. A substantial fraction of the cellular proteins make up a fibrous network, which seems to serve as a scaffold. This scaffold organises many other cell constituents and is concerned with both maintaining and altering the shape of the cell and with bringing about cell movement. Major components here are actin filaments, microtubules, and intermediate filaments. Intermediate filaments are stable compared with microtubules and actin filaments, which are continually assembling and disassembling. There is a large research industry devoted to analysing this process, particularly in relation to motility and the role of other proteins. There is also a major effort towards understanding how the cytoskeleton is linked to the plasma membrane. One view is that events at the surface may be transduced through the cytoskeleton to provide a signal to the cell nucleus. This is closely linked with the recent emphasis given to the possible importance of cell shape and the role of the extracellular matrix in determining cell behaviour.

Extracellular matrix and plasma membrane

Rather than being perceived as merely rather dull packing the extracellular matrix, which is secreted by the cells, is now seen as having a major role in determining cellular behaviour. This is particularly true of epithelia. For example, mammary gland epithelium in culture behaves quite differently depending on the nature of the matrix to which it is attached. This interaction may be mediated by the matrix determining cell shape. Again, the migration of cells is particularly dependent on the matrix. Major matrix components such as fibronectin have been identified only quite recently, and the classes of collagens seem to increase annually.

The plasma membrane is now considered in terms of a fluid bimolecular lipid membrane in which proteins, such as receptors, float, move, and interact. Attention is focused on transmembrane events mediated by these proteins. For example, some growth factors bind to receptors which activate a phospholipase on the inner surface. This splits phospholipid into inisitol triphosphate (ITP) and diacylglycerol (DAG). Inisitol triphosphate leads to an increase in calcium ions, whereas diacylglycerol activates protein kinase. The remainder of the pathway is complex.

It is remarkable that, with some exceptions like steroid hormones, we do not know the sequence of events that lead from an external signal to the cell's response. The sequence for glucocorticoid is well established: it enters the cell, binds a cytoplasmic protein, enters the nucleus, and attaches to specific sequences of DNA activating genes.[4] But in many other cases, including the action of insulin, growth hormones, and adrenaline, for which some of the early steps are known, the sequence peters out.

Cells are complex little organisms. This is particularly dramatically illustrated by oncogenes such as src. We know that the cell is transformed into a malignant cell by this gene—the sequence is fully known. But how does it do it? The gene codes for a protein kinase, but the sequence of cascade-like events is simply not known.

Cells are not what they used to be. Their study has been transformed by new techniques, particularly the analysis of cell components in molecular terms and the all persuasive influence of recombinant DNA technology. Remarkable progress has been made in decomposing and reassembling cellular structures and functions in vitro. A recent advance is the ability to reform nuclei around injected DNA in a cell-free extract. Perhaps the assembly of whole and new kinds of cells is not too far off. Will our grandchildren have 48 chromosomes?

1 Alberts B, Bray D, Lewis J, Raff M, Roberts K, Watson JD. *The molecular biology of the cell.* 2nd ed. London: Garland, 1989.
2 Palmiter RD, Brinster RL. Germ-line transformation of mice. *Annu Rev Genet* 1983;**20**:465–500.
3 Blau HM. How fixed is the differentiated state? Lessons from heterokaryons. *Trends Genet* 1989;**5**:268–72.
4 Yamamoto KR. Steroid receptor regulated transcription of specific genes and gene networks. *Annu Rev Genet* 1985;**19**:209–52.

Cell reproduction

Paul Nurse

The sequence of events and processes leading to cell reproduction is known as the cell cycle. During the cell cycle there is generally a duplication of cellular components followed by their partitioning into two daughter cells at cell division. For many components present in moderate to large amounts within the cell this duplication and partition need not be controlled exactly. The chromosomes, however, must be precisely replicated and segregated during each cell cycle to produce two viable daughter cells. Replication of the DNA making up the chromosomes occurs during S phase, and segregation of the chromosomes occurs during mitosis. These are the two major events of all cell cycles. In most cell cycles there are also gaps between these events: a G1 gap after mitosis and before S phase and a G2 gap after S phase and before mitosis.

When the cells are not actively undergoing reproduction they exit from the cell cycle and enter a quiescent state. This exit usually occurs during G1, and the quiescent state is called GO. Quiescent cells may be metabolically active, but they do not grow and do not undergo any of the events and processes of the cell cycle. The shift between the GO quiescent state and the cell cycle plays an important part in the growth and maintenance of the organism. Regulated entry into the cell cycle is necessary for proper development and for processes such as wound healing and tissue replenishment. If regulation becomes defective it may result in uncontrolled cell reproduction, leading to cancer.

I describe here some recent progress in our understanding of how the transition from quiescence to the cell cycle is controlled. I also consider briefly two further controls over cell reproduction which act during the cell cycle, one in late G1 before S phase and a second at the transition from G2 to mitosis.

Growth factors and their receptors

Mammalian cells may be grown in cell culture on the surface of a culture dish bathed in a nutrient medium supplemented with serum. If the cells are deprived of serum they stop dividing and become quiescent. Components called growth factors are the active principles present in serum that are necessary to stimulate entry into the cell cycle. Large numbers of different growth factors have now been isolated from various different tissues. They are generally proteins active at low concentrations—often less than 1 μg/ml can substitute for the serum. For maximum stimulation of cell growth, however, it may be necessary to add several different growth factors to the culture medium. Two commonly investigated growth factors are platelet derived growth factor (PDGF) and epidermal growth factor (EGF).

Growth factors stimulate cells to grow by binding to specific growth factor receptors located on the cell surface membrane. Many receptors have been found to be tyrosine protein kinases, which phosphorylate tyrosine residues of substrate proteins. The binding of a growth factor to its receptor is thought to activate the protein kinase activity, and this leads to phosphorylation of its protein substrates. How this stimulating signal is transduced to the rest of the cell is not well understood. In some cases signalling may entail breakdown of membrane inositol phospholipids thus generating second messengers, inositol triphosphate (ITP) and diacylglycerol (DAG). Coupling between this breakdown and the receptor may be brought about by guanosine triphosphate binding proteins regulating a phospholipase C enzyme that hydrolyses inositol phospholipid. Regardless of the precise mechanism of signal transduction, the consequence of growth factor binding is to generate a wide range of changes in the cell. Within minutes there are changes in concentrations of low molecular weight components, such as hydrogen and calcium ions and cyclic adenosine monophosphate (CAMP); later there are changes in specific transcript and protein concentrations. These changes result in an overall increase in the rates of synthesis of cellular protein and RNA, which leads to cell growth and eventual entry into the cell cycle.

Cell spreading and anchorage dependence

Cell spreading and attachment to a surface also influences whether cells enter the cell cycle. Normal cells not attached to a surface round up and become quiescent. On attachment to a surface they spread out and enter the cell cycle. This requirement to be attached to a surface is called anchorage dependence. Study of cells grown in a culture dish on special islands that restrain the extent of cellular spreading has shown that there is a correlation between the amount of spreading and the frequency with which a cell divides. When the island is large, cells continue dividing until they fill the island and form a confluent monolayer. Because the cells then have less room they round up and become quiescent. Wounding this monolayer by scraping away a strip of cells allows the cells at the edge of the wound to spread out and to re-enter the cell cycle.

Part of the explanation for these phenomena is likely to be that the growth factor receptors on the surface of spread out cells are more accessible to growth factors in the culture medium. There also seems to be an absolute requirement for cells to have some attachment to a surface. This may be necessary to generate changes in the cell cytoskeleton, which are required before cells can enter the cell cycle.

Cell transformation and oncogenes

Some cell lines do not show anchorage dependence and are not so dependent on the presence of growth factors in the culture medium. When grown on a culture dish surface they form layers of cells on top of each other instead of forming a confluent monolayer. Such cell lines are called transformed; they can be generated in various ways, such as by infection with certain animal viruses or after prolonged growth in culture. After injection back into an animal transformed cells often form tumours, which suggests that they are defective in the normal controls regulating entry into the cell cycle.

A class of genes called oncogenes has been found to be responsible for cell transformation. They have been isolated from transforming viruses and from DNA extracted from tumour cells. Homologues to oncogenes have been found in normal untransformed cells and are called proto-oncogenes. Cloning and sequencing of these genes have established that some of them encode previously identified components of the growth factor signal transducing pathway—for example, the proto-oncogene c-sis encodes a subunit of platelet derived growth factor, c-erbB a receptor for epidermal growth factor, and c-ras a guanosine triphosphate binding protein. Mutants of these proto-oncogenes are thought to generate oncogenes by altering the signal transducing pathway and stimulating entry into the cell cycle without the usual requirements for growth factors and anchorage dependence.

Relevance for the organism

These studies of cells in culture have provided a picture of how cell reproduction may be controlled in the intact organism. Cells remain quiescent unless they are exposed to particular growth factors and have the appropriate receptors to respond. Finer control is possible if several growth factors have to act together to elicit maximum response. Contacts between cells and surfaces can modulate the stimulation and consequently may be important for maintaining tissue cohesiveness and for preventing cells reproducing when detached from their current position. In these ways cells are stimulated to enter the cell cycle at the time and place necessary for the correct development and maintenance of the organism. Failure of these controls leads to cell transformation and possibly to the formation of tumours.

Commitment to the cell cycle

Once a cell has been stimulated to enter the cell cycle it remains subject to various controls. A major point of regulation occurs towards the end of G1, when commitment to the cell cycle takes place. In mammalian cells this is often known as the restriction point. Cells deprived of growth factors before this point will leave the cell cycle and enter the GO quiescent state. Once past this point, deprivation of growth factors often has little effect on the cell cycle in progress. Only after entering the G1 of the next cell cycle can the cell become quiescent. Before the restriction point is reached, cells are also more sensitive to inhibition of protein synthesis. If protein synthesis is inhibited to about half the normal level in such cells, they cannot proceed through to S phase and mitosis; once past the restriction point, however, inhibition to these levels has no effect on the ability of cells to complete that cell cycle.

A similar commitment control, called start, acts in the G1 phase of the unicellular eukaryotic yeasts. Several genes required for yeast cells to complete start have been identified. One of these, cdc2, has been shown to encode a threonine and serine protein kinase. Once past start the cells begin the programme of events which lead eventually to S phase and mitosis. For example, after start there is synthesis of specific components which will be required for DNA replication during S phase.

Transition from G2 to mitosis

A further point of control occurs at the initiation of mitosis. Fusion of mammalian cells at different stages of the cell cycle has shown the existence of factors that can induce cells into mitosis. These factors can be assayed direct by injecting cell extracts into frog oocytes arrested at meiotic nuclear division. If the extracts contain these factors then the oocyte is induced into meiosis. Such experiments have shown that the inducing factors are present in mitotic mammalian cells and vary in concentration during cleavage of the frog embryo, the concentration reaching a peak as the cell enters mitosis.

Genes that are important for initiating mitosis have also been identified in yeast. Interestingly, the same gene (cdc2) that is implicated in cell cycle commitment also plays a part in the initiation of mitosis. The activity of the cdc2 protein kinase peaks during mitosis. This activation requires association with another protein called cyclin, but the timing of activation and thus of mitotic onset is determined by dephosphorylation of a key tyrosine in the cdc2 protein. The kinase phosphorylates several proteins that have roles during mitosis.

Similarities between yeast and mammals

Human cells have been found to contain a cdc2 gene homologue. The gene can substitute for the yeast gene and encodes a structurally closely related protein kinase, which is regulated in a similar way to the equivalent protein in yeast. The conservation of a cell cycle control gene in yeasts and humans indicates that aspects of the control will probably be similar in all eukaryotic organisms. It also suggests that protein phosphorylation has a key role in mammalian cells, both in commitment to the cell cycle and at the initiation of mitosis. Rapid advances may now be expected in this subject because yeast is convenient for such studies and may be used as a surrogate organism for investigating the mode of action and regulation of the human cdc2 homologue.

Further reading

Several excellent chapters relevant to cell reproduction may be found in Bradshaw RA, Prentis S, eds. *Oncogenes and growth factors*. Amsterdam: Elsevier Science Publications, 1987: Bishop JM. Trends in oncogenes, 1–10; Weinberg RA. Cellular oncogenes, 11–6; Burgess A. Growth factors and oncogenes, 123–34; Hunter T. Oncogenes and growth control, 135–42; Majerus PW, Wilson DB, Connolly TM, Bross TE, Neufeld EJ. Phosphoinositide in turnover provides a link in stimulus-response coupling, 269–74; Nurse P. Cell cycle control genes in yeast, 275–83.

O'Neill CH, Jordan P, Ireland G. Evidence for two distinct mechanisms of anchorage stimulation in freshly explanted and 3T3 Swiss mouse fibroblasts. *Cell* 1986;**44**:489–96.

Lee MG, Nurse P. Complementation used to clone a human homologue of the fission yeast cell cycle control gene cdc2. *Nature* 1987;**327**:31–5.

Kirschner M, Newport J, Gerhart J. The timing of early developmental trends in Xenopus. *Trends in Genetics* 1985;**1**:41–7.

Nurse P. Universal control mechanism regulating onset of M-phase. *Nature* 1990;**344**;503–9.

The cell nucleus

R A Laskey

The cell nucleus is the information centre of the cell. It is responsible for copying highly selected regions of the genome into RNA and for supplying precisely regulated amounts of specific RNA molecules to the cytoplasm, where they are translated into proteins. In addition it must duplicate its entire structure every time the cell divides.

This chapter considers how the nucleus is organised to perform the immensely complex task of selective information retrieval. Several different experimental approaches have elucidated how DNA is organised in the nucleus and how specific regions of DNA are selected for expression at particular times and in particular cells of the body. One of the key problems in selective information retrieval in the nucleus becomes obvious if we consider how densely DNA is packed into the nucleus. Each chromosome consists of an individual double helix of DNA about 40 mm long by 2 nm wide packed into a nucleus of about 6 μm in diameter. The implications of these dimensions can be seen more clearly from a simple scale model enlarged one million times. On this scale the DNA in each chromosome would resemble thin string, 2 mm in diameter but 40 km long, and the total DNA in one nucleus would reach from London to Naples. Yet on the same scale this 2000 km of DNA must all be packaged into a nucleus of only 6 m diameter in such a way that specific regions must remain fully accessible for highly regulated expression.

The answer to how such a densely packed structure can function in selective information retrieval lies in a precise three dimensional nuclear architecture. The contents of the nucleus are not just a randomly mobile solution but are highly organised into a hierarchy of ordered structures.

Chromatin and chromosomes

The fundamental unit of DNA packaging in higher organisms is the nucleosome, in which 146 base pairs of DNA are wrapped twice around an octamer of histones, consisting of two each of histones H2A, H2B, H3, and H4. These small basic proteins neutralise the acidic charges of DNA and they also shorten it by providing spools on which the DNA is coiled.

A fifth histone, called H1, occupies the site at which DNA enters and leaves the nucleosome. When H1 is present nucleosomes can coil into a solenoid with six nucleosomes in each turn. This has two important consequences. Firstly, coiling compacts DNA so that it can be divided between progeny cells at mitosis, and, secondly, it allows selected regions of DNA to be folded away so that only a specific subset of the genetic information is available for expression in a particular type of cell. Active genes are packaged into a different, and more accessible, chromatin conformation from that of inactive genes in the same cell. Thus nucleosomes allow a structural level of gene regulation by permitting selective access to regions of DNA. In addition, the absence of a nucleosome or a small group of nucleosomes from the region of DNA just upstream of a gene can be important in gene regulation. These regions are known as DNase 1 hypersensitive sites because of their extreme susceptibility to nuclease cleavage. They frequently correspond to regions of DNA that are essential for correct gene regulation (see below).

Not only is the DNA in each chromosome packaged into nucleosomes but it is also organised into a series of loops, which extend radially from an axial scaffold. Loops vary in size but on average they contain 20 000–100 000 base pairs each. Specific proteins bind to and grip DNA sequences that define the bases of these loops. This level of organisation persists when chromosomes are condensed for cell division or when they are decondensed for transcription and replication during interphase. Furthermore, the organisation does not seem to change when cells differentiate. Instead it seems to define a basal level of organisation within the nucleus (fig 1).

Each chromosome consists of a linear chain of genes, but it also contains other structural features, including centromeres and telomeres. Centromeres attach chromosomes to the spindle during mitosis and meiosis, whereas telomeres form the termini of chromosomes. Both centromeres and telomeres have been isolated and characterised by selecting for these functions on artificial chromosomes in yeast. Centromeres allow plasmids to segregate stably to progeny cells during mitosis and meiosis, whereas telomeres allow

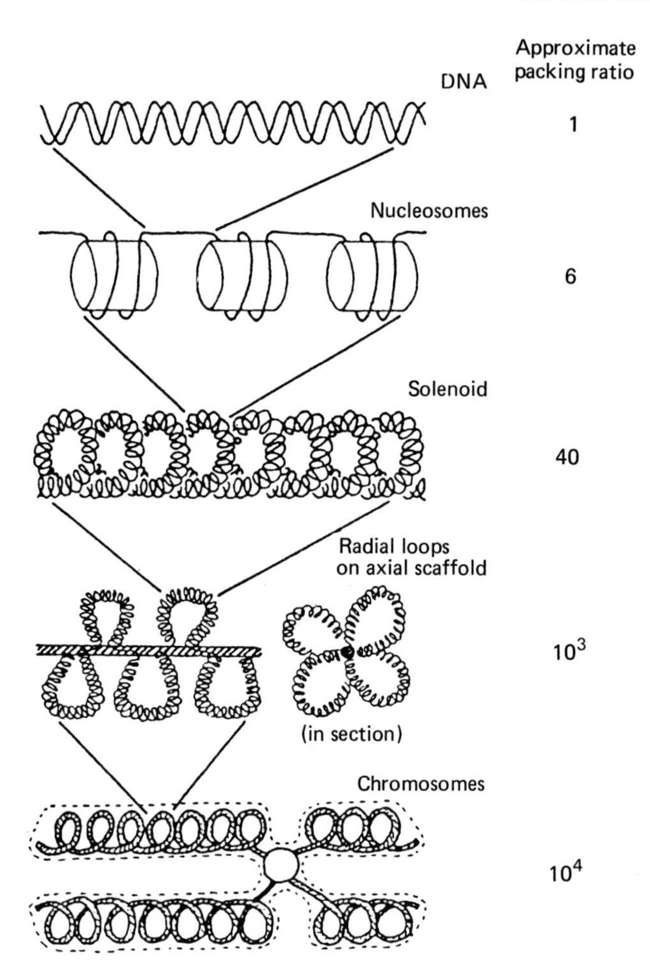

FIG 1—Summary of structural models for the packing of DNA into metaphase chromosomes. (Reprinted with permission from Laskey, 1986).

linear DNA molecules to complete their replication correctly. DNA polymerases are unable to replicate right to the ends of a linear chromosome, so without a specialised telomere, chromosomes would become slightly shorter each cell generation. The telomeric DNA sequence is a simple repeat, which can fold back to form a hairpin loop or even a four stranded structure involving hydrogen bonding between four deoxyguanosine residues to form "G quartets". These simple repeats are not synthesised by copying the parental DNA strands, but by an enzyme called telomerase, which contains a short RNA template. Telomerase copies this RNA template into DNA, extending the ends of the chromosomes.

The ability to construct artificial chromosomes in yeast has allowed components essential for chromosome function to be identified. Apart from centromeres, telomeres, and genes, yeast chromosomes also require specific DNA sequences to serve as initiation sites for DNA synthesis (origins of replication). Obviously we would like to know more about equivalent sequences from mammalian cells, and an active search is in progress. Human telomeres have been isolated and they are strikingly similar to those of yeast, but human centromeres and replication origins have been more elusive.

Gene expression and its control

The opportunity to clone genes and to reintroduce them into intact cells has permitted rapid advances in our understanding of gene expression and its regulation. It is possible to delete regions of DNA from within or around genes and to reclone the truncated genes in order to assay their expression. In this way the DNA sequences responsible for regulated expression of a wide range of genes have been identified.

In many cases regulatory sequences have been mapped to the flanking DNA which lies just upstream of the transcription start, though there are important exceptions in which regulatory information lies downstream or even within the transcribed region of a gene. Some upstream flanking sequences are found to be conserved between many genes whereas others confer tissue specificity. For example, the upstream regulatory region of the metallothionein gene contains a widely conserved TATA box, about 30 nucleotides before the transcription start, but it also contains three specific regulatory sequences, two of which mediate induction by heavy metals and one of which confers regulation by glucocorticoids. This third site binds the hormone receptor and thus confers hormonal control.

Many other proteins that act as specific transcription factors have now been discovered. They bind to regulatory sequences adjacent to specific genes and enable those genes to be transcribed. In general they occur at about 10 000 copies per cell, an observation that poses an immediate problem. If each gene needed 10 000 copies of its own specific transcription factor, the concentration of

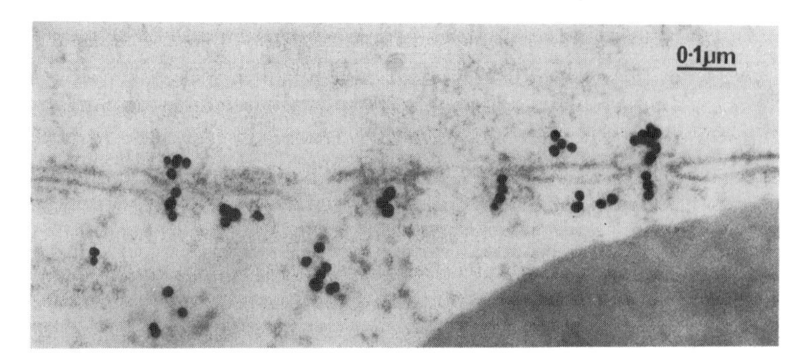

FIG 2—Electron micrograph of a section through the nuclear envelope of a frog oocyte that has been injected with colloidal gold particles coated with the nuclear protein "nucleoplasmin". Gold particles can be seen aligned through nuclear pore complexes. (Courtesy of A D Mills).

factors within the nucleus would be enormous. This seems to be partly solved through the use of different combinations of shared factors. This may be particularly common in "housekeeping" genes, which are required to be active in all types of cell. Thus combinations of relatively few factors can regulate a much larger number of genes, just as combinations of the 26 letters of the alphabet can specify a vast number of words.

Once an activating combination of transcription factors has assembled on a gene's regulatory region, it activates the basal transcription machinery, including the appropriate RNA polymerase, to synthesise a messenger RNA precursor. This is usually processed by polyadenylation of its 3′ end and splicing to remove "introns" from within the RNA sequence before it is exported to the cytoplasm as mature messenger RNA (mRNA). These post-transcriptional processing steps are performed by various small nuclear ribonucleoprotein (snRNP) particles. The ability to perform these reactions in vitro has led to advances in our understanding of their mechanisms.

The role of the nuclear envelope in traffic control

The nucleus is surrounded by two complete layers of membrane, but these do not exclude small molecules because the membranes are perforated at frequent intervals by nuclear pore complexes. These are grommet like structures that permit the free passage of small molecules up to the size of small proteins, but they appear to regulate selectively the transit of larger molecules in both

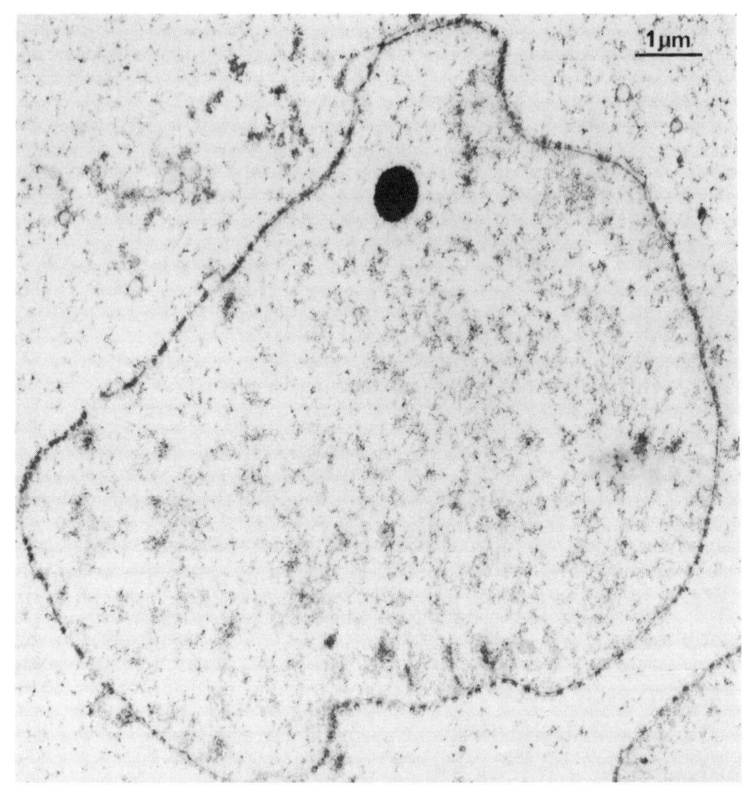

FIG 3—Electron micrograph of a section through a nucleus reconstituted from demembranated frog sperm chromatin by incubation in a cell free system that also assembles nuclei from naked DNA. (See Blow and Laskey (1986) for details. Photograph by courtesy of A D Mills).

directions. Thus above about 50 kilodaltons molecular weight only "nuclear" proteins can enter the nucleus, and other proteins are excluded.

The information that specifies selective entry resides in a very short stretch of amino acids. In the case of SV40 T antigen this sequence is: proline, lysine, lysine, lysine, arginine, lysine, valine. In many other cases, such as the acidic protein nucleoplasmin, two short clusters of basic amino acids are required, separated by a spacer. When a short synthetic peptide of such a sequence is cross linked to cytoplasmic proteins it causes them to accumulate in the nucleus. The nuclear pore has been identified as the route of protein entry by coating colloidal gold particles with a nuclear protein and using electron microscopy to follow their fate after injection into cytoplasm. The gold particles can be seen aligned through the centre of the nuclear pore in figure 2.

The nuclear pore is also known to be the route of export of ribonucleoprotein particles out of the nucleus, including ribosomes and messenger RNA. This process is also highly selective, resulting in the export of mature transcripts but the retention in the nucleus of immature transcripts.

Rebuilding the cell nucleus

There have been substantial advances in our ability to reconstruct "nuclei" from condensed chromatin or even purified DNA in cell free systems. Figure 3 shows a large fully enveloped nucleus that has been formed from demembranated frog sperm chromatin in a cell free system derived from frog eggs. Similar structures have now been formed from purified DNA, including DNA from bacterial viruses and plasmids. Not only is such DNA assembled into chromatin; it also becomes surrounded by nuclear membranes containing functional nuclear pores. Thus these pseudonuclei fill up with nuclear proteins and even replicate their DNA in vitro under cell cycle control.

There is growing evidence that the structural organisation of the nucleus has important roles in DNA replication and mRNA production. The opportunity to reconstruct functioning nuclei and functioning chromosomes opens up new opportunities to understand how these complex structures perform their extraordinary functions.

Further Reading

General

Alberts B, Bray D, Lewis J, Raff M, Roberts K, Watson JD. *Molecular biology of the cell.* 2nd ed. New York and London: Garland, 1989: chapters 9 and 10.

Chromatin and chromosomes

Jackson DA. Structure-function relationships in eukaryotic nuclei. *Bioessays* 1991;**13**:1–10.

Price CM. Centromeres and telomeres. *Current Opinion in Cell Biology* 1992;**4**:379–84.

Smith MM. Histone structure and function. *Current Opinion in Cell Biology* 1991;**3**:429–37.

Gene expression

Carey M. Mechanistic advances in eukaryotic gene activation. *Current Opinion in Cell Biology* 1991;**3**:452–60.

Mitchell PJ, Tjian R. Transcriptional regulation in mammalian cells by sequence-specific DNA binding proteins. *Science* 1989;**245**:371–8.
Ptashne M, Gann AAF. Activators and targets. *Nature* 1990;**346**:329–31.
Lamond AI. Nuclear RNA processing. *Current Opinion in Cell Biology* 1991;**3**:493–501.

Import and export

Silver PA. How proteins enter the nucleus. *Cell* 1991;**64**:489–97.
Nigg EA, Baeuerle PA, Lührmann R. Nuclear import-export: in search of signals and mechanisms. *Cell* 1991;**66**:15–22.

Rebuilding the cell nucleus

Blow JJ, Laskey RA. Initiation of DNA replication in nuclei and purified DNA by a cell-free extract of Xenopus eggs. *Cell* 1986;**47**:577–87.
Laskey RA. Prospects for reassembling the cell nucleus. *J Cell Sci (Suppl)* 1986;**4**:1–9.
Laskey RA, Leno GH. Assembly of the cell nucleus. *Trends Genet* 1990;**6**:406–10.

Stem cells in normal growth and disease

B R Clark, T M Dexter

Each day, under normal conditions, the regenerating tissues of the adult human lose more than 100 g of mature cells.[1,2] To maintain the integrity of these tissues in response to this constant cell loss, the mature cells must be replaced. The regenerating tissues (for example, gastrointestinal epithelium, bone marrow, skin, testes) also have the capacity to produce extra cells in response to conditions of additional cell loss caused by bleeding, infection, or injury. Mature cells produced by these systems, however, are highly specialised cells and are usually incapable of further growth. The numbers of mature cells are maintained by the proliferation and development of more primitive cells, known as stem cells. The process by which stem cells develop into their clearly distinguishable mature cell counterparts is termed differentiation. Stem cells are found in most tissues, although clearly they are not acting in every case to renew the tissue constantly. Some tissues (for example, those of the liver) possess an impressive ability to repair damage, which indicates the presence of a latent stem cell population that is induced to proliferate and differentiate under stress conditions.

Origin of stem cells

Two models have been proposed to explain the persistence of stem cells throughout life.[3-5] The first model suggests that a fixed number of primitive cells is laid down during embryogenesis to supply the body's needs throughout its lifetime. Stem cells are recruited into proliferation, differentiation, and development as required—like the recruitment of oocytes. As these stem cells proliferate they progressively lose their ability to produce more stem cells or to act as founder cells of the various mature cell lineages—in other words, the stem cell pool declines with age. The second model also suggests that a small population of stem cells arises during embyronic development but that these cells can reproduce themselves (undergo self renewal) to produce daughter cells, which retain the same proliferative and developmental potential as the original parental cells. In this model, therefore, the recruitment of stem cells into proliferation and development does not necessarily lead to a reduction in the number of stem cells. The weight of evidence strongly supports this second model.[6 7] The critical point here is that life and death processes must be balanced—exit from the stem cell population as a consequence of differentiation or death must be balanced by an input of cells into the stem cell population. How, then, are tissues organised to maintain not only this basal balance but also the ability to respond in stress conditions? Many lessons have been learned from the haemopoietic system, and this will be used as a primary example to discuss aspects of stem cell biology and how this knowledge can be applied for therapeutic advantage.

Stem cells: capable of sustained self renewal

Haemopoietic stem cells arise in the yolk sac and then migrate to the fetal liver and subsequently to the developing bones.[8] Within the bone marrow cavities the stem cells associate with the complex range of cells and extracellular matrix of the bone marrow stroma.[9] Extensive work has shown that the different kinds of blood cells—neutrophils, monocytes and macrophages, eosinophils, basophils, erythrocytes, megakaryocytes (platelets), osteoclasts, B lymphocytes, and T lymphocytes—are all derived from a common stem cell pool in the marrow.[10–13] In other words, a haemopoietic stem cell is multipotent—capable of giving rise to multiple haemopoietic lineages. In other tissues, the presence of such a spectacularly multipotent stem cell is not quite so obvious. In the testes, for example, only spermatogonia are generated by sperm stem cells, and in the skin the epithelial stem cells give rise to mature keratinocytes only. However, stem cells in all tissues share a common feature—the rate of stem cell self renewal must be balanced against differentiation and mature cell formation and loss. It is this ability of stem cells to persist by self renewal that is common to all stem cells.[14]

Organisation of regenerating tissues

How do the stem cells "know" when to divide and generate cells capable of giving rise to mature cells to maintain the complement of cells in a tissue? How do stem cells maintain a constant stem cell pool by self renewal? The answers lie in the organisation of tissues.

Intestinal epithelium covers the surface of the intestinal tract. Intestinal epithelial stem cells are present in the crypts of Lieberkühn—a set of glands at the bases of the intestinal villi. Epithelial cells are generated in the crypts and migrate on to the villi to replace the mature epithelial cells that are shed or sloughed off. When crypts are examined in cross section using techniques to label the DNA of dividing cells, it becomes clear that the stem cells in the crypt—localised in specific sites—give rise to more differentiated progeny, which have a limited ability to undergo self renewal.[15] These daughter cells expand in number and differentiate while migrating out of the crypt and on to the villi. In the final stages, the mature cells lose the ability to divide, forming most of the epithelium covering the villi. Thus, in the intestine, the largest contribution to mature cell numbers occurs through the expansion of the population of differentiating cells with a limited ability to self renew—the progenitor cell population (figure 1).

Progenitor cells are found in other systems, such as the skin or bone marrow, where cell numbers are amplified. In haemopoiesis, for example, 3×10^{11}—that is 300 000 million—cells are required each day.[2] It only requires 10 divisions of a single progenitor cell and its daughter cells to give rise to over 1000 cells (2^{10}, figure 1). It is the expansion of the progenitor population—and not the stem cells—that gives rise to the cellular amplification seen in several tissues.

Nevertheless, why are stem cell pools not depleted as they generate more progenitor cells throughout a lifetime? The most popular hypothesis to account for this suggests that there exists a finite number of sites in tissues where stem cells can maintain their primitive phenotype—because of a complex requirement for accessory cells and extracellular matrix.[5] In tissues such as the gastrointestinal epithelium and skin, stem cells are localised to discrete regions of the tissue.[15 16] For the purpose of this discussion, the term "niche" will be used to describe the finite, privileged environment that maintains any stem cell in a primitive state. Niches act to limit and maintain the size of the stem cell pool, ensuring that it is not doubled after stem cell self renewal. Only one of the

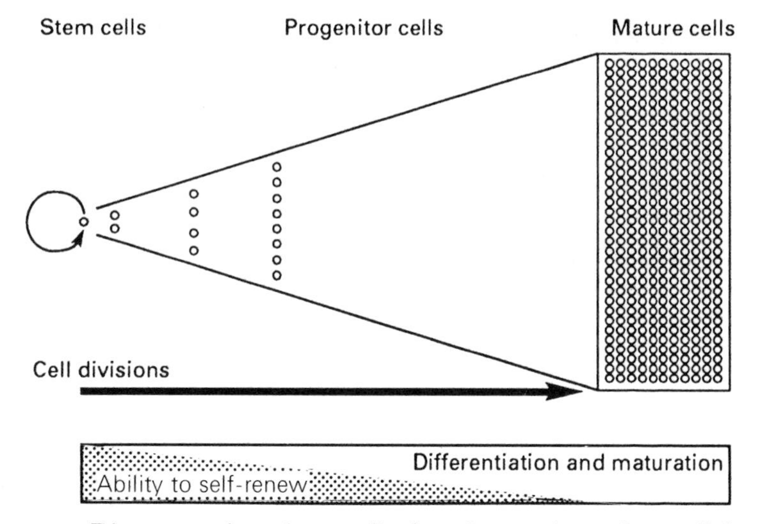

FIG 1—Diagram to show the contribution of progenitor cells to cellular amplification. Progenitor cells and their progeny undergo division to give rise to the mature cell populations. This ability to self renew declines as the cells undergo differentiation into mature cells. For clarity, not all division steps are shown.

daughter cells can remain in the privileged niche, where it preserves its capacity to survive and self renew. The other daughter cell moves out of the niche to differentiate or die.

Indeed, cell death is a common phenomenon in many tissues, where it may be used to regulate the numbers of stem cells, progenitor cells, and mature cells.[17 18] In this instance, cell death is not simply a result of the cell wasting away through lack of nutrition or anchorage. This regulatory cell death appears to be an active decision to die swiftly when an essential supportive stimulus is removed, or when the cell receives a signal instructing it to die.[18] This process of active cell death is termed apoptosis. The swift destruction of a cell undergoing death by apoptosis—as opposed to death by wasting—is caused by the start of a "self destruct" sequence of events, which destroys the cell and packages the pieces in such a way that they can be scavenged efficiently.[19] Apoptosis ensures that cells with a capacity to self renew do so in a controlled manner—under the influence of the appropriate signals—and that cells in the wrong place or surviving at the wrong time are promptly and efficiently destroyed. Many examples of apoptosis are seen in tissue destruction and remodelling. Perhaps one of the most obvious is seen in the generation of fingers from fetal limb paddles, when interdigital cells are lost in a controlled manner to give a five fingered hand.[17]

Systemic and local regulation of haemopoiesis

In the bone marrow, there is good evidence of spatial organisation of haemopoietic stem cells and progenitor cells as well as of a finite number of stem cell niches in the bone marrow stroma.[9 20] Thus, in this respect, maintainance of the haemopoietic stem cell pool is thought to be analogous to the maintainance of stem cell populations in tissues such as the gut or skin.

How is the haemopoietic system regulated to generate the required numbers and types of mature cells? In the haemopoietic system, events such as phlebotomy or coagulation decrease the number of circulating mature cells and result in compensatory changes within the bone marrow, which act to normalise the levels of circulating cells. The number of circulating cells can also be increased in response to, for example, infection or a decrease in atmospheric pressure. Clearly, haemopoiesis is influenced by a wide variety of systemic factors that can act upon haemopoietic progenitor cells in the bone marrow.

Perhaps the most striking example of systematic regulators in haemopoiesis involves the production of mature red blood cells (RBCs). It has been known for nearly 100 years that the body produces a hormone, erythropoietin, which can stimulate the generation of red blood cells.[21 22] Erythropoietin is a glycoprotein hormone generated by kidney cells in response to changes in the local oxygen tension. Erythropoietin is essential for the development of red blood cells from erythroid progenitors in the marrow. In the absence of erythropoietin, erythroid progenitors are thought to die by apoptosis. Thus the supply of erythropoietin regulates the production of mature erythrocytes.

In addition to erythropoietin many other haemopoietic growth factors (HGFs) have been discovered.[23] The genes encoding these glycoproteins have been isolated, permitting the production of sufficient pure haemopoietic growth factors for clinical applications (table I). The populations of haemopoietic progenitor cells that these factors influence can be identified through the use of in vitro culture assays. Under appropriate culture conditions, individual haemopoietic progenitor cells divide and give rise to colonies of cells, which may include more differentiated haemopoietic

TABLE I—*Some (by no means all) examples of growth factors, negative regulators, and differentiation inducing agents*

Growth factor	Some biological actions	Some clinical applications
G-CSF	Stimulation of neutrophil progenitors, neutrophil production, and function of mature neutrophils	Preventing neutropenia after chemotherapy, bone marrow transplantation, or in conditions of congenital or acquired neutropenia, eg cyclic neutropenia, AIDS
GM-CSF	Stimulation of neutrophil progenitors, neutrophil production	As G-CSF
Erythropoietin	Stimulation of erythropoiesis	Treating anaemia of chronic renal disease
M-CSF	Stimulation of myelopoiesis	Enhancing recovery of myeloid cells after bone marrow transplantation
SCF	Acts on very primitive haemopoietic cells to enhance responses to other growth factors	Enhancing response to other growth factors
IL2	Stimulates T lymphocytes, influences capillary endothelium	Lymphokine activated killer (LAK) cell treatment of malignancies. Response seen with IL2 alone in renal cell cancer
IL3	Acts with other growth factors to stimulate haemopoietic cell division	In combination with other factors, possibly improves number and range of mature cells generated
IL4	Stimulates B lymphocyte cell division	Possibly a useful immunomodulator?
IL6	Stimulates B lymphocyte cell division. Putative autocrine growth factor for myeloma cells. Acts in concert with other factors on primitive haemopoietic cells	Possibly enhances the action of other growth factors. Perhaps useful to stimulate myeloma cells into cell cycle before chemotherapy
IFN alfa	Interferons act on cell division and differentiation. Discovered as compounds with in vitro antiviral activity	Decreasing Philadelphia chromosome positive cells in chronic myeloid leukaemia (CML), inducing remission in hairy cell leukaemia
IFN gamma		Some activity on haematological and solid malignancies
TGF β	Decreases cell cycling in primitive progenitor cells in bone marrow and gut	Protecting gut and marrow stem cells during cycle-specific chemotherapy
MIP 1α	Decreases cell cycling in primitive bone marrow cells	Protecting gut and marrow stem cells during cycle-specific chemotherapy

TABLE I—*Growth factors, negative regulators, differentiation inducing agents (continued)*

Growth factor	Some biological actions	Some clinical applications
LIF/DIA	Inhibits cycling of leukaemic cells in vitro. Maintains primitive phenotype of embryonal stem cells	Possibly in stimulating platelet production
Retinoic acid	Influences differentiation of cell types possessing suitable receptor	Treating some acute promyelocytic leukaemias, oral leukoplakia, acne, and squamous cell carcinoma
EGF	Stimulates formation of skin	Treating partial thickness skin wounds including burns

G-CSF = granulocyte colony stimulating factor, GM-CSF = granulocyte and macrophage colony stimulating factor, M-CSF = macrophage colony stimulating factor, SCF = stem cell factor (or c-kit ligand, mast cell growth factor), IL = interleukin, IFN = interferon, TGF = transforming growth factor, MIP = macrophage inflammatory protein, LIF/DIA = leukaemic inhibitory factor/differentiation inhibitory activity, EGF = epidermal growth factor.

cells. A cell that gives rise to a colony is termed a colony forming cell (CFC), and the type of cells found within the mature colony indicates the type of colony forming cell. Some of the colonies contain cells of multiple lineages—for example, erythroid cells, neutrophils, and megakaryocytes—and are obviously derived from a multipotent progenitor cell. Other colonies contain mature cells representative of only one or two cell lineages and are derived from unipotent or bipotent progenitor cells. For example, a colony containing granulocytes and macrophages is generated by a bipotent granulocyte and macrophage colony forming cell (GM-CFC). Some haemopoietic growth factors are termed colony stimulating factors (CSFs) because of their action upon colony forming cells in these in vitro assays. For example, granulocyte and macrophage colony stimulating factor (GM-CSF) supports the formation of colonies by progenitor cells capable of differentiation into granulocytes and macrophages.

Based on the in vitro colony forming data, colony forming cells can be placed into a scheme of haemopoiesis within which multipotent and bipotent progenitor cells become committed to a single lineage (figure 2). These lineage committed progenitors and their daughter cells have a limited ability to self renew, but they can divide to give rise to the cellular amplification seen in haemopoiesis.

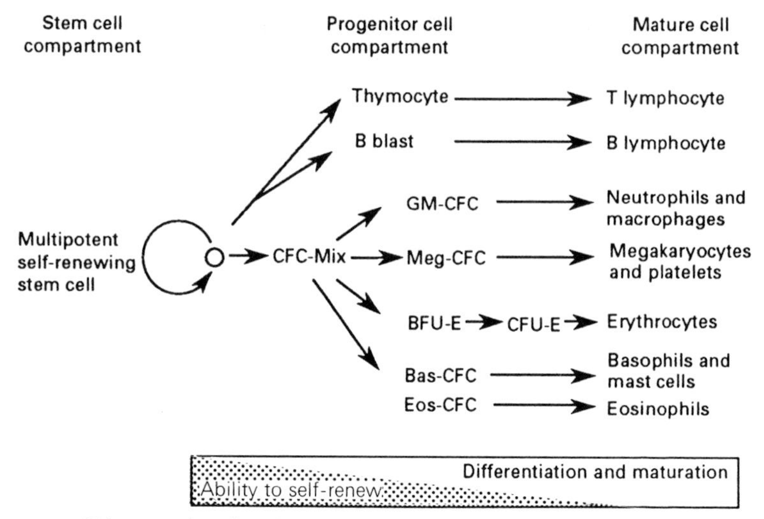

FIG 2—Diagram showing the differentiation and development of stem cells in a haemopoietic hierarchy. (CFC-Mix = a multipotential cell producing mixed myeloid colonies in vitro, GM-CFC = granulocyte and macrophage colony forming cell, BFU-E = a primitive erythroid progenitor cell, CFU-E = a more mature erythroid colony forming cell, Eos-CFC = eosinophil colony forming cell, Bas-CFC = basophil and mast cell progenitor, Meg-CFC = megakaryocyte colony forming cell.)

Because it has proved extremely difficult to detect many of these soluble haemopoietic growth factors in the serum or the bone marrow under normal conditions, it is not yet clear what factors are responsible for maintaining basal haemopoiesis in vivo. However, many haemopoietic growth factors are thought to be produced within the marrow by cells of the bone marrow stroma. Some haemopoietic growth factors synthesised by stromal cells are produced as transmembrane glycoproteins and are found on the stromal cell surface.[24 25] Other haemopoietic growth factors are produced as soluble glycoproteins but may become sequestered by components of the bone marrow extracellular matrix.[26 27] Both of these localised, non-diffusing sources of haemopoietic growth factors are available to the developing haemopoietic cells.[24 26 27] Thus the cells and extracellular matrix of the bone marrow form a complex microenvironment for haemopoietic cells. However, this local control of haemopoiesis can be overridden by haemopoietic growth factors that are produced in response to stress conditions. For example, haemopoietic growth factors such as granulocyte colony stimulating factor (G-CSF) and granulocyte and macrophage colony stimulating factor (GM-CSF) are produced at sites of an immune response to infection.[28] These haemopoietic growth factors act on the bone marrow to promote the generation and release of neutrophils. One additional feature of many haemopoietic growth factors is that they also act to stimulate the function of mature cells. For example, granulocyte colony stimulating factor and granulocyte and macrophage colony stimulating factor enhance neutrophil metabolism and adhesion.[29–32] Thus growth factors produced at sites of infection may enhance the function of neutrophils in that region as well as promoting the production of more neutrophils and their migration to inflammatory sites.

In examples examined so far, cells such as erythrocytes or gastrointestinal epithelium are produced in response to cell loss. Feedback mechanisms—such as the detection of anoxia or perhaps the availability of space to migrate—act on progenitor cells to regulate the production of mature cells. In response to infection or wounding, stimulatory factors ensure the rapid production and migration of additional cells. However, in haemopoiesis several populations of cells exist that are very small in number and are generally quiescent. Lymphocytes, for example, constitute around 0·05% of circulating cells. Thus, unlike erythrocytes, a shortage of the normally quiescent lymphocytes may not be detected immediately and may only become apparent by a secondary means—for example, infection.

How are the numbers of lymphocytes regulated? As in erythroid development, apoptosis has a role in basal lymphopoiesis. More than the daily requirement of developing T and B lymphocytes is generated. The numbers of both cell types are regulated by the lymphoid organs—the thymus for T lymphocytes and the spleen and secondary lymphoid organs for B lymphocytes. Some 95–99% of all developing T lymphocytes produced in the thymus (thymocytes) are thought to receive a "death" signal and die by apoptosis.[33] Only thymocytes that recognise foreign antigens—and not self antigens—survive and proliferate. The T lymphocytes leaving the thymus survive for many months or years. On the other hand, B lymphocytes have a much shorter half life—weeks to months—and must be renewed constantly.[34] Newly produced, virgin B lymphocytes leaving the marrow are thought to have been selected already, at least in part, for the ability to respond to foreign antigens only. The challenge for a virgin B lymphocyte is to receive a "survival signal" in the spleen or a secondary lymphoid organ.[35]

Failure to receive such a signal is thought to result in death by apoptosis. Virgin B lymphocytes compete with the recirculating B lymphocyte pool for these signals. If the mature B lymphocyte pool is depleted then more of the virgin B lymphocytes released from the marrow can receive a survival signal. Thus, by using competition at the level of the lymphoid organs, the recruitment of new mature B lymphocytes is directly linked to the size of the existing pool.

Stem cells in disease

Dysregulation of stem cells and their more differentiated progeny is thought to contribute to several clinical conditions. The balance between cell production and cell loss that is normally tightly regulated in tissues can be subverted at any point from the level of stem cell self renewal to the inappropriate survival of maturing cells (table II).

In benign neoplastic conditions, many so called pre-malignant lesions exhibit some degree of escape from the normal mechanisms that govern tissue turnover. For example, warts or intestinal polyps are formed when cells divide under the influence of abnormal growth regulation. These pre-malignant events can pave the way for critical mutations to genes responsible for growth regulation, which results in cellular transformation into a malignant neoplasm (see chapter entitled Genes and cancer).

Several genes that promote cellular division have been identified. Mutations or inappropriate expression of these genes are thought to be able to stimulate cells to divide. Some of these genes are called transforming oncogenes because of their suspected involvement in promoting malignant events.[36] Another set of genes acts to restrict cell division. Some are termed anti-oncogenes or tumour suppressor genes,[37] and in several malignancies tumour suppressor genes are lost or mutated. In vitro experiments have shown that in such cases the addition of the appropriate tumour suppressor gene(s) to cells from tumours results in a loss of a malignant phenotype, which suggests that the loss of tumour suppressor function contributed to the malignant transformation. A combination of mutations of both transforming oncogene and tumour suppressor gene is thought to be involved in the progression of polyposis to gastrointestinal carcinoma.[38]

Many clinical conditions are the result of dysregulated cell production (table II). As the technology is not (yet) available to

TABLE II—*Some examples of conditions (many pre-malignant) caused by dysregulated growth in a range of tissue types*

Condition	Comment
Myelodysplastic syndrome	Defective haemopoietic cell production by stem cell pool. Haemopoiesis commonly clonal—the product of a single stem cell. Pre-malignant condition, evidence for transformation to acute myeloid leukaemia
Secondary polycythemia	Overproduction of erythrocytes due to excess levels of erythropoietin.
Polyposis coli	Premalignant condition that predisposes to colorectal carcinoma.
Warts	Dysregulated production of epithelial cells caused by papillomaviruses. Papillomaviruses and genital warts are linked to increased risk of cervical carcinoma.
Lentigo maligna	Defect of melanocytogenesis predisposing to malignant melanoma.
Tylosis	Rare condition of hyperkeratosis of palms and soles, predisposes to oesophogeal carcinoma.
Some small cell lymphomas	The bcl genes, thought to be involved in suppressing apoptosis, are involved in chromosome translocations. Whereas the developing B lymphocytes are thought to decrease bcl expression and become susceptible to apoptosis, B lymphocytes with altered bcl gene expression survive in the lymph node in the absence of a "survival signal".

alter genetic lesions within cells in vivo, present attempts to influence the growth of cells rely on our knowledge of how these systems can be modulated by external stimuli. As discussed, dividing cells can be influenced by the surrounding environment—the presence of soluble factors that are adjacent to the cell. A wide range of soluble factors is known to act on tissues to influence aspects of cellular division or differentiation.[23 39] Some of these factors, termed growth factors, are increasingly used in the management of conditions in which cellular regulation is dysfunctional or the body is unable to produce sufficient mature cells.

The therapeutic use of growth factors

Several growth factors, especially the haemopoietic growth factors, are in regular clinical use (table I). These factors have roles in modifying the generation and function of mature cells. Haemopoietic growth factors are applied to stimulate the production of mature cells when endogenous growth factors are reduced (for example, erythropoietin in chronic renal disease[40]) or when the haemopoietic system has been compromised—after bone marrow transplantation or damage by chemotherapy or radiation.[41] By expanding the number of progenitor cells and supporting differentiation into mature cells, haemopoietic growth factors enable compromised bone marrow to produce more mature cells to prevent infection or anaemia. In addition to promoting the generation of cells, growth factors can induce cells to begin the process of cell division. The cells undergoing division may be targeted by chemotherapy, enabling drugs to "hit" quiescent malignant cells that would otherwise escape treatment.[42]

In contrast, factors that act to suppress cellular division—termed negative growth regulators—may also have clinical applications (table I).[43–46] For example, transforming growth factor β (TGF β) can inhibit the cell division of primitive progenitor cells in the gastrointestinal epithelium and bone marrow. These populations of cells are often damaged by radiation or chemotherapy used in cancer treatment to kill dividing cells. Suppressing cell division in the gastrointestinal epithelial or bone marrow progenitors can possibly reduce side effects and overcome some of the dose limiting toxicity, which permits treatment with larger doses. This approach will only work if the malignant cell population does not respond to the negative regulator and continues to divide. Some leukaemic cells do not respond to negative regulators such as TGF β or MIP 1α (macrophage inflammatory protein 1α). Thus, MIP 1α may suppress the division of normal haemopoietic progenitor cells whereas leukaemic cells will continue to divide. Giving chemotherapy that kills dividing cells allows the leukaemic cells to be selectively targeted and the normal cells protected.[47]

The abnormal production of cells may be rectified in the future by using agents that promote differentiation. Immature cells that are undergoing self renewal may be induced to differentiate, giving rise to mature cells incapable of further division. This principle is currently in use against some dysplastic cutaneous lesions in which retinoic acid can act as a differentiating agent.[48,49] This general approach may be applicable to other diseases, such as acute myeloid leukaemias. In acute promyelocytic leukaemias, a translocation of chromosomes 15 and 17 places the gene for the retinoic acid receptor next to a region that directs its expression.[50] These leukaemias are very responsive to treatment with retinoic acid.[51]

The specificity of action of many growth factors makes them suited to tasks involving the modification of abnormal cellular growth and function. Future approaches to modifying cell growth and differentiation may be found in new techniques such as targeted gene treatment. Such an approach would permit the permanent correction of several genetic defects by modifying tissues at the level of a stem cell.

1 International Commission on Radiobiological Protection. Report on the Task Group on Reference Man. Oxford: Pergamon, 1975:144. (ICRP publication No 23.)
2 Cronkite EP, Feinendegen LE. Notions about human stem cells. *Blood Cells* 1976;**2**:263–84.
3 Hellman S, Botnick LE. Stem cell depletion: an explanation of the late effects of cytotoxins. *Int J Radiat Biol* 1977;**2**:181–4.
4 Rosendaal M, Hodgson GS, Bradley TR. Organisation of haemopoietic stem cells: the generation-age hypothesis. *Cell Tissue Kinet* 1979;**12**:17–30.
5 Schofield R. The relationship between the haemopoietic stem cell and the spleen colony forming cell: a hypothesis. *Blood Cells* 1978;**4**:7–25.
6 Ross EAM, Anderson N, Micklem HS. Serial depletion and regeneration of the murine hematopoietic system: implications for hematopoietic organisation and the study of aging. *J Exp Med* 1982;**155**:432–44.
7 Harrison DE, Astle CM. Loss of stem cell repopulating ability upon transplantation. *J Exp Med* 1982;**156**:1767–79.
8 Metcalf D, Moore MAS. *Haemopoietic cells*. Amsterdam: North Holland Publishing Co, 1971.
9 Allen TD, Dexter TM, Simmons PJ. Marrow biology and stem cells. In: Dexter TM, Testa NG, Garland J, eds. *Colony stimulating factors*. New York: Marcel Dekker, 1990:1–38.
10 Abramson S, Miller RG, Phillips RA. The identification in adult bone marrow of pluripotent and restricted stem cells of the myeloid and lymphoid lineages. *J Exp Med* 1979;**145**:1567–79.
11 Snodgrass R, Keller G. Clonal fluctuation within the haematopoietic system of mice reconstituted with retrovirus infected stem cells. *EMBO J* 1987;**6**:3955–60.
12 Till JE, McCulloch EA. Haemopoietic stem cell differentiation. *Biochim Biophys Acta* 1980;**605**:431–59.
13 Lemischka IR, Raulet DH, Mulligan RC. Developmental potential and dynamic behaviour of haemopoietic stem cells. *Cell* 1986;**45**:917–27.
14 Lajtha LG. Stem cell concepts. In: Potten CS, ed. *Stem cells*. London: Churchill Livingstone, 1983:1–11.
15 Potten CS, Hendry JH. Stem cells in the small intestine. In: Potten CS, ed. *Stem cells*. London: Churchill Livingstone, 1983;155–99.
16 Potten CS. Stem cells in epidermis from the back of the mouse. In: Potten CS, ed. *Stem cells*. London: Churchill Livingstone, 1983;200–32.
17 Duvall E, Wylie AH. Death and the cell. *Immunol Today* 1986;7:115–9.
18 Williams GT. Programmed cell death: apoptosis and oncogenesis. *Cell* 1991;**65**:1097–8.
19 Savill J, Dransfield I, Hogg N, Haslett C. Vitronectin receptor-mediated phagocytosis of cells undergoing apoptosis. *Nature* 1990;**343**:170–3.
20 Lord BI. Cellular and architectural factors influencing the proliferation of hematopoietic stem cells. In: *Differentiation of normal and neoplastic hematopoietic cells*. Cold Spring Harbor, NY, USA: Cold Spring Harbor Laboratory, 1978:775–88.
21 Carnot P. Sur le mechanisme d'hyperglobulie provoquée par le serum d'animaux en renovation sanguine. *C R Acad Sci [III]* 1906;**111**:344–6.

22 Krumdiek N. Erythropoietic substance in the serum of anemic animals. *Proc Soc Exp Biol Med* 1943;**54**:14–7.
23 Metcalf D. The molecular control of cell division, differentiation commitment and maturation in haemopoietic cells. *Nature* 1989;**339**:27–30.
24 Anderson DM, Lyman SD, Baird A, Wignall JM, Eisman J, Rauch C, *et al.* Molecular cloning of mast cell growth factor, a hematopoietin that is active in both membrane and soluble forms. *Cell* 1990;**63**:235–43.
25 Rettenmier CW, Rousse MF, Ashmun RA, Ralph P, Price K, Sherr CJ. Synthesis of membrane-bound colony-stimulating factor 1 (CSF-1) and down modulation of CSF-1 receptors in NIH3T3 cells transformed by cotransfection of human CSF-1 and c-fms (CSF-1 receptor). *Mol Cell Biol* 1987;7:2378–87.
26 Gordon MY, Riley GP, Watt SM, Greaves MF. Compartmentalisation of a hemopoietic growth factor (GM-CSF) by glycosoaminoglycans in the bone marrow micro-environment. *Nature* 1987;**326**:403–5.
27 Roberts RA, Gallagher JT, Spooncer E, Allen TD, Bloomfield F, Dexter TM. Heparan sulphate-bound growth factors: a mechanism for stromal cell-mediated haemopoiesis. *Nature* 1988;**332**:376–8.
28 Metcalf D. *The molecular control of blood cells*. Cambridge, Mass, USA: Harvard University Press, 1988.
29 Yuo A, Kitagawa S, Ohsaka A, Saito M, Takaku F. Stimulation and priming of human neutrophils by granulocyte-colony stimulating factor and granulocyte-macrophage colony stimulating factor: qualitative and quantitative differences. *Biochem Biophys Res Commun* 1990; **171**:491–7.
30 Balazovich KJ, Almeida HI, Boxer LA, Recombinant human G-CSF and GM-CSF prime human neutrophils for superoxide production through different signal transduction mechanisms. *J Lab Clin Med* 1991;**118**:576–84.
31 Aranout MA, Wang EA, Clark SC, Sieff CA. Human recombinant granulocyte-macrophage colony-stimulating factor increases cell-to-cell adhesion and surface expression of adhesion promoting surface glycoproteins on mature granulocytes. *J Clin Invest* 1986;**78**:597–601.
32 Weisbart RH, Kwan L, Golde DW, Gasson JC. Human GM-CSF primes neutrophils for enhanced oxidative metabolism in response to the major physiological chemoattractants. *Blood* 1987;**69**:18–21.
33 Goldstein P, Ojcius DM, Young JD. Cell death and the immune system. *Immunol Rev* 1991;**121**:29–65.
34 Gray D, Skarvall H. B Cell memory is short lived in the absence of antigen. *Nature* 1988;**336**:70–3.
35 Leanderson T, Källberg E, Gray D. Expansion, selection and mutation of antigen-specific B cells in germinal centres. *Immunol Rev* 1992;**126**:47–61.
36 Bodmer W. Somatic cell genetics and cancer. *Cancer Surv* 1988;7:239–50.
37 Baker SJ, Markowitz S, Fearon ER, Wilson JK, Vogelstein B. Suppression of human colorectal carcinoma cell growth by wild type p53. *Science* 1990;**249**:912–5.
38 Kinzler KW, Nilbert MC, Su-LK, Vogelstein B, Bryan TM, Levy DB, *et al.* Identification of FAP locus genes from chromosome 5q21. *Science* 1991;**253**:661–5.
39 Cross M, Dexter TM. Growth factors in development, transformation and tumourigenesis. *Cell* 1991;**6**:271–80.
40 Sundal E, Businger J, Kappler A. Treatment of transfusion-dependent anaemia of chronic renal failure with recombinant erythropoietin. A European multi-centre study in 142 patients to define dose regimen and safely profile. *Nephrol Dial Transplant* 1991;**6**:955–65.
41 Sheridan WP, Lorstyn G, Wolf M, Dodds A, Lusk J, Maher D, *et al.* Granulocyte colony-stimulating factors and neutrophil recovery after high dose chemotherapy and autologous bone marrow transplantation. *Lancet* 1989;**ii**:891–5.

42 De Witte T, Muus P, Haanen C, Van der Lely N, Koekman E, Van der Locht A, *et al.* GM-CSF enhances sensitivity of leukemic clonogenic cells to long term low dose cytosine arabinoside with sparing of normal clonogenic cell. *Behring Inst Mitt* 1988;**83**:301–7.
43 Lord BI, Dexter TM, Clements JM, Hunter MA, Gearing AJH. Macrophage inflammatory protein protects multipotent haemopoietic cells from the cytotoxic effects of hydroxyurea in vivo. *Blood* 1992;**79**:59–63.
44 Keller JR, Ellingsworth LR, McNiece IK, Quesenberry PJ, Sing GK, Ruscetti FW. Transforming growth factor β directly regulates primitive murine hematopoietic cell proliferation. *Blood* 1990;**75**:596–602.
45 Keller JR, Sing GK, Ellingsworth LR, Ruscetti FW. Transforming growth factor β: possible roles in the regulation of normal and leukemic cell growth. *J Cell Biochem* 1989;**39**:175–84.
46 Ruscetti FW, Dubois C, Falk LA, Jacobsen SE, Sing G, Longo DL, *et al.* In vivo and in vitro effects of TGF-β1 on normal and neoplastic haemopoiesis. In: Bock GR, Marsh J, eds. *Clinical applications of TGF-β*. Chichester: Wiley, 1991;212–31. (Ciba Foundation Symposium No 157.)
47 Tsyrlova IG, Lord BI. Inhibitor of CFU-S proliferation preserves normal haemopoiesis from cytotoxic drug in long-term bone marrow culture—L1210 leukaemia model. *Leuk Res* 1989;**13** (suppl 1):40.
48 Lippman SM, Parkinson DR, Itri LM, Weber RS, Schantz SP, Ota DM, *et al.* 13-Cis-retinoic acid and interferon alpha 2a: effective combination therapy for advanced squamous cell carcinoma of the skin. *J Natl Cancer Inst* 1992;**84**:235–41.
49 Edwards L, Jaffe P. The effect of topical tretinoin on dysplastic nevi. A preliminary trial. *Arch Dermatol* 1990;**126**:494–9.
50 Kakizuka A, Miller WH, Umesono K, Warrell RP, Frankel SR, Murty VV, *et al.* Chromosomal translocation t(15;17) in human acute promyelocytic leukemia fuses RAR alpha with a novel putative transcription factor, PML. *Cell* 1991;**66**:663–74.
51 Castaigne S, Chomienne C, Daniel MT, Ballerini P, Berger R, Fenaux P, *et al.* All-trans retinoic acid as a differentiation therapy for acute promyelocytic leukemia. I. Clinical results. *Blood* 1990;**76**:1704–9.

Sorting signals and cellular membranes

Graham Warren

Cells use a surface receptor protein to trap and internalise the cholesterol they need for membrane synthesis. The receptor protein is synthesised on ribosomes, which become bound to the membrane of the endoplasmic reticulum (fig 1). The receptor protein is then transported by vesicles through the Golgi apparatus, where it undergoes various modifications before appearing on the cell surface. Here the receptor protein binds circulating low density lipoprotein (LDL), a cholesterol carrier, and the two are internalised by surface invaginations called coated pits. These pinch off to form coated vesicles, and uncoating is followed by fusion with endosomes. Low density lipoprotein is parted from the receptor, delivered to lysosomes, and degraded; the cholesterol so released passes to the endoplasmic reticulum, where membrane synthesis takes place. The receptor is recycled from the endosomes back to the cell surface, where it participates in roughly a hundred more rounds of internalisation. It too is then degraded in lysosomes.

Cellular compartments

The life cycle of the low density lipoprotein receptor is characterised by movement from one membrane bound compartment to the next in a highly ordered and efficient manner. Compartmentation is a major feature distinguishing eukaryotes from prokaryotes and enables them to carry out many different cellular functions under optimum conditions. As the focus of this review is proteins I shall define a compartment as a collection of soluble proteins surrounded by a membrane. Different compartments have different collections of soluble and membrane proteins,

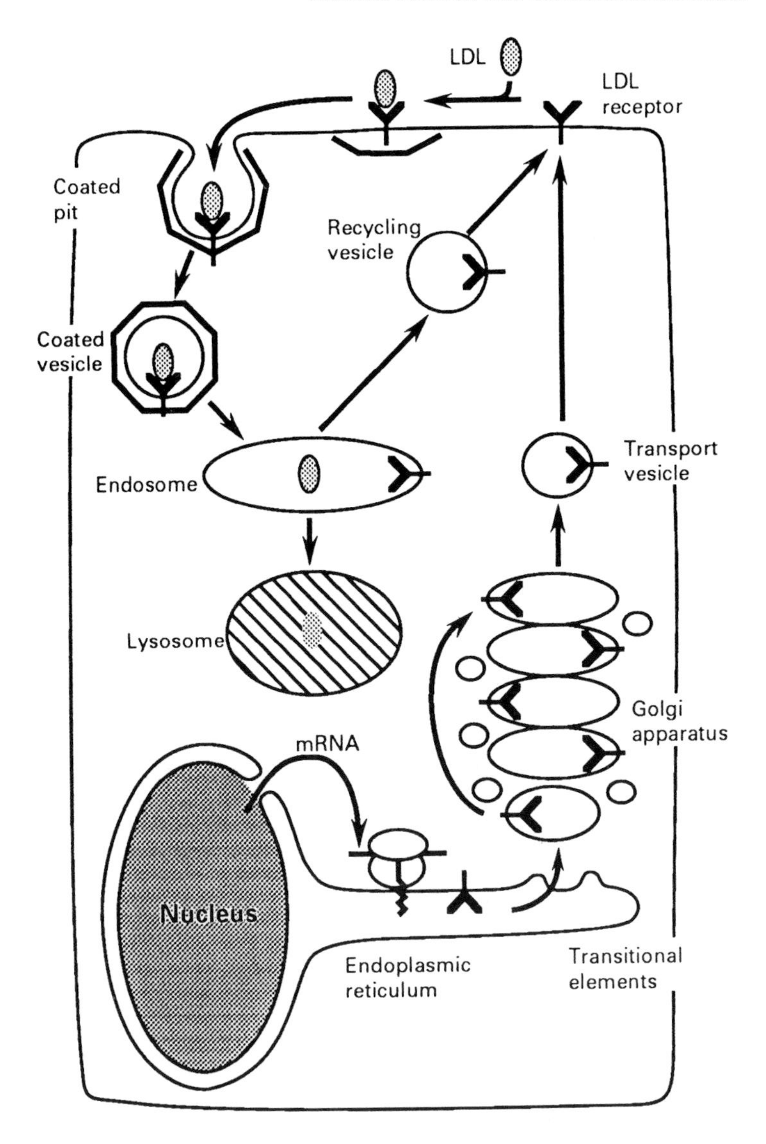

FIG 1—Synthesis and cycling of the low density lipoprotein receptor. Messenger ribonucleic acid (mRNA) coding for the receptor leaves the nucleus and is translated on ribosomes, which become attached to the membrane of the endoplasmic reticulum. After transport to the cell surface the receptor binds low density lipoprotein (LDL) and delivers it to the endosomes, recycling back to the cell surface to bind more low density lipoprotein. Degradation of low density lipoprotein in lysosomes releases the cholesterol needed for membrane synthesis.

though some proteins may be common to both compartments. The function of the endosome, for example, is to separate complexes such as low density lipoprotein from their receptor. Binding of the two is sensitive to acid so the membrane of the endosome contains proton pumps that keep the inside acidic, causing the low density lipoprotein to dissociate from the receptor. The lysosome also has proton pumps in the membrane to keep the inside acidic, but this is to allow the soluble degradative enzymes that it contains to work optimally. One might ask why there is a need for both endosomes and lysosomes. As lysosomes are acidic, low density lipoprotein could be both released from the receptor and then degraded. The answer is that the receptor would also be degraded. The endosome serves to protect the receptor for further rounds of internalisation.

The sorting problem

What is true for endosomes and lysosomes is also true for all other cellular compartments. They have each evolved to carry out a particular set of functions that confers unique advantages on the eukaryotic cell. These specialisations, however, create a major problem, which is the focus of this review. The problem is in two parts, the first concerning the growth and division of eukaryotic cells. These must be accompanied by the growth and division of the membrane bound compartments, which in turn necessitates the synthesis of the membrane and soluble proteins that make up each compartment. The largest compartment in the cell, the cytoplasm (bounded by the plasma membrane), is specialised for protein synthesis probably because most of the proteins synthesised in the cell are for use in the cytoplasm. But this specialisation means that the proteins needed for all of the compartments illustrated in fig 1 must also be synthesised in the cytoplasm. In other words, they are synthesised at a site different from the site at which they carry out their cellular function. This explains why the low density lipoprotein receptor is not inserted direct into the plasma membrane; synthesis starts in the cytoplasm, and the endoplasmic reticulum is specialised for the assembly of compartmental proteins. This also means that the low density lipoprotein receptor is synthesised alongside proteins destined for other compartments. Therefore, the mechanisms that transport them to their correct location in the cell must also separate them en route from proteins destined for other compartments.

The second part of the problem arises once the protein has reached its destination. Proteins such as the low density lipoprotein receptor can function only if they move from compartment to compartment. Low density lipoprotein receptors can enter endosomes, but there are many plasma membrane proteins that function at the cell surface and spend their lifetime there. They, unlike the low density lipoprotein receptor, must be prevented from entering coated pits. Similarly, during receptor recycling, endosomal proteins must be prevented from entering the recycling vesicles. Both parts of this problem together constitute what is referred to as the sorting problem; perhaps the easiest way to think about it over all is as a problem of protein purification. Each compartment must maintain its identity and does so by regulating the proteins that enter and leave. Proteins that are allowed to enter must be purified away from proteins that are not allowed to enter. Similarly, proteins that are allowed to leave must be freed from those that make up the compartment's collection of proteins. The endoplasmic reticulum, for example, does not allow cytoplasmic proteins to cross the membrane and prevents endoplasmic reticulum proteins from leaving when exported proteins such as the low density lipoprotein receptor leave. How does it do this? Obviously, the mechanism must be able to distinguish between proteins, which implies that they are tagged in such a way as to identify their correct location in the cell. I should now like to show, with the examples of the low density lipoprotein receptor, a lysosomal enzyme, a Golgi protein, and a soluble protein of the endoplasmic reticulum, what little we know about these tags and the mechanisms that recognise them.

Sorting signals

Synthesis begins on free ribosomes and differs from the synthesis of cytoplasmic proteins in that the first 20 or so amino acids constitute a signal specifying the endoplasmic reticulum as the first destination (fig 2). This signal is interpreted by a signal recognition particle, which binds to the signal and temporarily halts synthesis of the protein. The ribosomal complex then docks with a docking protein, found only on the endoplasmic reticulum, ensuring that further synthesis is coupled to transfer across the membrane of the endoplasmic reticulum and no other. The soluble endoplasmic reticulum protein and the lysosomal enzyme pass completely across the membrane and end up freely soluble inside; the low

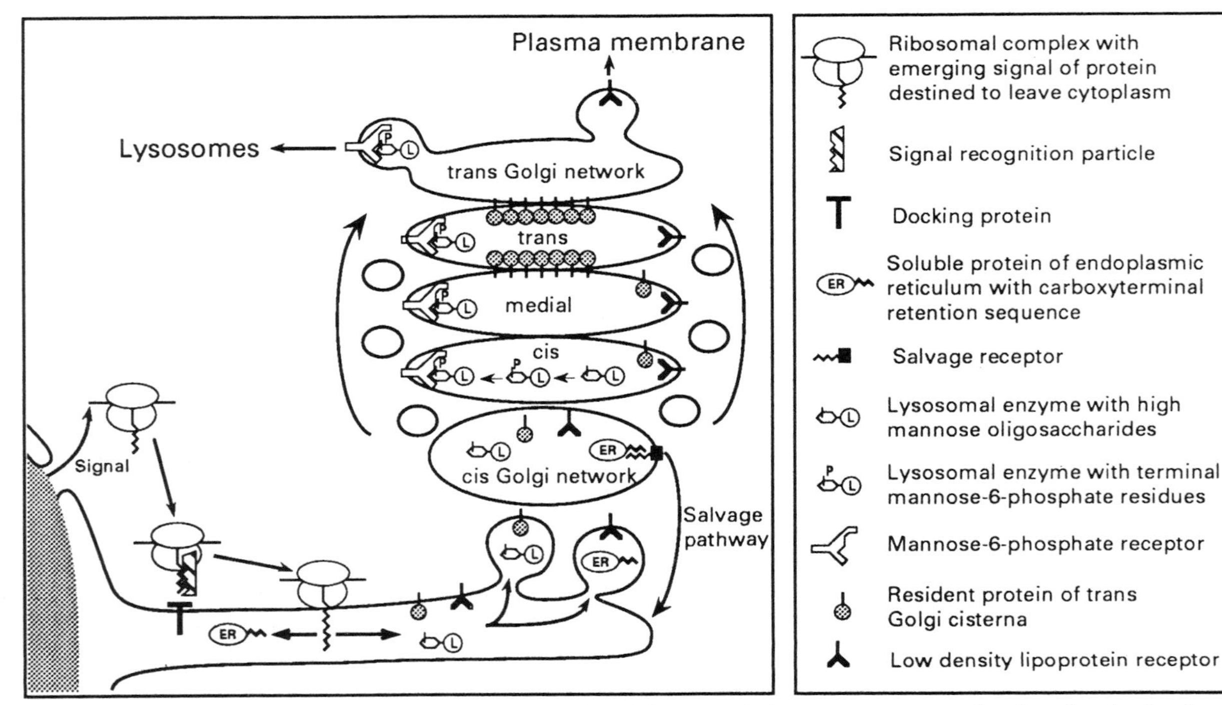

FIG 2—Synthesis and transport of proteins destined for different cellular compartments. Sorting signals that have been characterised include those that enable a protein to leave the cytoplasm, those that retain it in the endoplasmic reticulum or the Golgi apparatus, and those that direct it to lysosomes.

density lipoprotein receptor and the Golgi protein stop part of the way across, ending up as membrane proteins. As each protein crosses the membrane the signal sequence is removed, its job done, and oligosaccharides containing many mannose residues are added to the protein. These oligosaccharides have considerable importance for the subsequent sorting of the lysosomal enzyme.

These newly synthesised proteins are then transported to their correct destinations. The route to the cell surface is known as the default pathway because it does not require any sorting signals. Movement from the endoplasmic reticulum to the Golgi, from compartment to compartment within the Golgi, and from there to the cell surface is a constitutive process, mediated by small vesicles shuttling between the compartments. Newly synthesised proteins destined for locations other than the cell surface require one of three types of signal: retrieval signals, which permit recovery of proteins that have been inadvertently lost from the compartment in which they normally function; retention signals, which stop a protein moving more than a certain distance along the default pathway; and diversion signals, which move a protein on to a different pathway.

Retrieval signals are found on soluble proteins within the lumen of the endoplasmic reticulum. Some are assembly proteins, which help to fold up newly synthesised proteins. Because they are soluble they can get caught up in the flow of exported proteins and so are, occasionally, lost. When they arrive in the cis Golgi network they are salvaged by a specific receptor, which returns them (fig 2). This receptor recognises escaped proteins because they all carry a tag at the C-terminus, an invariant sequence of four amino acids (lysine, aspartic acid, glutamic acid, and leucine).

Retention signals are found on Golgi proteins, which must stop part way along the default pathway. Examples include those enzymes responsible for modifying the oligosaccharides bound to transported proteins. The signal, unexpectedly, is contained within the region of the protein that spans the lipid bilayer. This region is thought to catalyse oligomerisation, leading to large protein aggregates that are unable to move any further (fig 2).

The lysosomal enzyme has a diversion signal. When it arrives in the cis cisterna it is recognised by specific enzymes that add phosphate to the mannose residues, generating a mannose-6-phosphate tag. This is recognised by a receptor, also present in the cis cisterna, which then carries the enzyme by default through the medial and trans cisternas until it reaches the trans Golgi network.

Diversion occurs at this point because the receptor with bound enzyme is packaged into vesicles different from those carrying proteins to the cell surface (fig 2). Once the enzyme reaches the lysosomes the mannose-6-phosphate tag is hydrolysed. This releases it from the receptor, which recycles back to the cis Golgi cisterna for further rounds of lysosomal protein transport.

The low density lipoprotein receptor moves by the default pathway to the cell surface and so has no need of signals to get it there. Once there, however, it begins to cycle and to do that it requires information that causes it to be trapped selectively in coated pits. The tag for this selective trapping resides in that part of the receptor exposed to the cytoplasm. Mutations in this part yield receptors that are expressed on the cell surface and bind low density lipoprotein, but they are not efficiently internalised. Cholesterol cannot be taken up, and children with this defect suffer from familial hypercholesterolaemia. The tag is extremely efficient, allowing a 5000-fold purification of the receptor in a single step. This ensures that only those proteins that are meant to go to the endosomes are delivered there. All others are excluded from coated pits. The mechanism that carries out this selection is still unknown.

In summary we should remember that the three dimensional organisation of the cell demands two types of information: information to transfer the protein to that part of the cell where it is to function, and information that allows it to carry out its function. In the simplest case this entails merely keeping it in a particular compartment. In the more complicated case information is needed to move it specifically between compartments. Future work will uncover more of the signals that encode this information and assess the nature of the mechanisms that decode it.

General references

Goldstein JL, Brown MS, Anderson RGW, Russell DW, Schneider WJ. Receptor-mediated endocytosis. *Annu Rev Cell Biol* 1985;**1**:1–40.

Kornfeld R, Kornfeld S. Assembly of asparagine-linked oligosaccharides. *Annu Rev Biochem* 1985;**54**:631–64.

Machamer CE. Golgi retention signals: do membranes hold the key? *Trends in Cell Biology* 1991;**1**:141–4.

Pearse B, Bretscher MS. Membrane recycling by coated vesicles. *Annu Rev Biochem* 1981;**50**:85–101.

Pelham HR. Control of protein exit from the endoplasmic reticulum. *Annu Rev Cell Biol* 1989;**5**:1–23.

Pfeiffer SR, Rothman JE. Biosynthetic protein transport and sorting by the endoplasmic reticulum and Golgi. *Annu Rev Biochem* 1987;**58**:829–52.

How do receptors at the cell surface transmit signals to the cell interior?

Robert H Michell

The cells of the body are bathed in fluids whose major constituents—for example, the ions and proteins of plasma—are held at remarkably constant concentrations over long periods. These extracellular fluids also contain, at very low and constantly changing concentrations, an extraordinary number of chemically diverse molecules that work together to control cell and tissue behaviour and thus integrate body function (table I). Some of the signals, notably neurotransmitters, are released from one cell and recognised by another within a very limited space: examples include acetylcholine acting within the confines of a neuromuscular junction and various stimulatory and inhibitory neurotransmitters released at synapses between neurones. By contrast, classic hormones are disseminated throughout the body (or at least throughout the extraneural tissues), where they initiate tissue specific responses in those cells that bear receptors that recognise them.

Between these two extremes lie many other extracellular signals whose actions are localised but directly affect much larger volumes of tissue than classic neurotransmitters at close synapses. Such agents include (a) peptide neurotransmitters that, before their destruction, may diffuse away from a nerve ending and act on cells a few tens of micrometres away; (b) hypothalamic (neuro)hormones that are carried to the anterior pituitary (their major target organ) by a specialised local circulation; (c) the prostaglandins (prostacyclin from endothelium and thromboxane A_2 from activated platelets), subendothelial collagen, and ADP (secreted from platelets) that regulate platelet behaviour during incipient thrombus formation at the site of blood vessel injury; (d) the platelet

TABLE I—*Examples to illustrate the variety of extracellular stimuli to which mammalian cells can respond*

Light
Mechanical or osmotic perturbation
Smells and tastes, chemoattractants
Interactions between cells (eg sperm with egg, lymphocyte with antigen presenting cell)
Neurotransmitters (amines, amino acids, and small peptides)
(Neuro)Hormones (small amines, small peptides, proteins)
Cytokines, colony stimulating factors, growth factors, and cell differentiation and survival regulators
Antigens—either intact or fragments presented on self major histocompatibility complex
Fatty acid metabolites (prostanoids etc)
Inflammatory mediators (eg histamine, platelet activating factor, complement peptides, tachykinins)

derived and other polypeptide growth factors that control the repair of these and other wounds; and (e) histamine, leukotrienes, platelet activating factor, lymphokines, and other mediators of inflammatory events. Even the responses of your nose to an interesting smell and of an egg to the first contact of a fertilising sperm seem to be little different biochemically from the responses of major organs to hormones and neurotransmitters.

So how do the cells of the body, each of which carries an identical genome that has the potential to encode maybe 100 000 proteins, manage to mount correct responses to such an astonishing range of external chemical influences (table I)? The answer, which has emerged from research in many laboratories over the past 30 years or so, is remarkably simple and elegant. Basically, each extracellular stimulus has its own unique receptor (or receptors) but the onward transmission of information into cells channels the information from these multifarious receptors through only a limited number of signalling pathways that are built into the plasma membrane. Thus the molecular components of any one signalling pathway can allow a plethora of extracellular agents to control a wide variety of intracellular, and often tissue specific, response systems in the appropriate target cells. The principles of this economical mode of transmission of information are summarised in figure 1.

Most extracellular stimuli are from molecules that are large or hydrophilic, or both, and they interact promptly and with high affinity with receptors on the cell surface, but there are notable exceptions to this general description. These include steroid

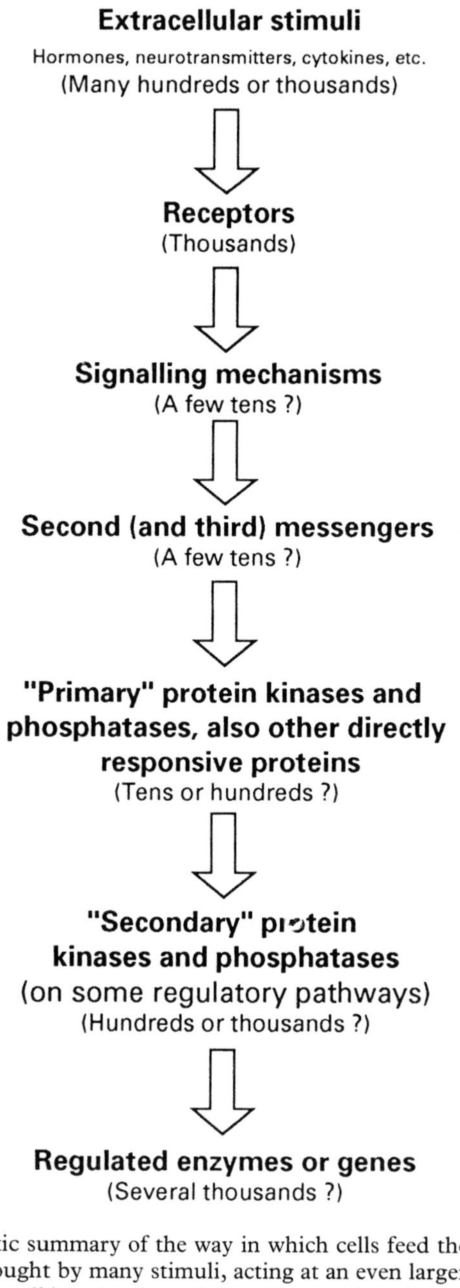

Extracellular stimuli

Hormones, neurotransmitters, cytokines, etc.
(Many hundreds or thousands)

Receptors
(Thousands)

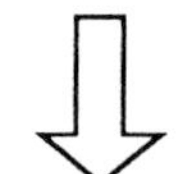

Signalling mechanisms
(A few tens ?)

Second (and third) messengers
(A few tens ?)

"Primary" protein kinases and phosphatases, also other directly responsive proteins
(Tens or hundreds ?)

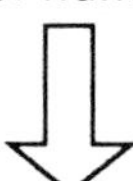

"Secondary" protein kinases and phosphatases
(on some regulatory pathways)
(Hundreds or thousands ?)

Regulated enzymes or genes
(Several thousands ?)

FIG 1—Schematic summary of the way in which cells feed the regulatory information brought by many stimuli, acting at an even larger number of receptors, to the cell interior through a smaller number of signal transduction pathways.

hormones, thyroid hormones, and retinoids, and also those hypolipidaemic agents that drive peroxisome proliferation in liver and other tissues (for example, benzofibrate). These lipid soluble molecules enter cells and bind to intracellular receptor proteins, and the resulting complexes of hormones and receptors then interact direct with chromatin to control the expression of genes: they will not be considered further in this review.

Recognition of extracellular stimuli

The exquisitely selective sites at the cell surface that recognise extracellular stimuli and even discriminate between closely related molecules (such as the nonapeptide neurohormones, oxytocin and vasopressin) were named "receptors" long before we learned that they are membrane proteins firmly integrated within the structure of the plasma membrane. In addition, those early studies sometimes disclosed striking heterogeneities in the measurable pharmacological characteristics of tissue responses to single agents, so leading to the suggestion that one type of molecule can often influence cells by interacting with more than one type of receptor: these hypothetical entities were named, for example, the α_1, α_2, and β receptors for catecholamines, the muscarinic and nicotinic receptors for acetylcholine, and the H_1 and H_2 receptors for histamine. Such multiple receptors, with their proved or potential applications as drug targets, have more recently been recognised for a large number of extracellular stimuli—for example, dopamine, serotonin, and vasopressin—and the even more recent molecular cloning of an enormous number of receptor proteins has shown that the true receptor multiplicity is greater than had ever been suggested by classic pharmacological studies.

Thus the number of individual receptor proteins encoded within the genome of every nucleated cell is much greater than the number of different chemical signals that is recognised by any of the cells of the body at any time during our lives—maybe a few thousand genes, representing a small percentage of the genome. However, at any one time during our lives only a relatively few of these genes that encode receptor proteins are turned on in any individal type of differentiated cell, so that each cell expresses receptors for and responds to only that limited range of stimuli for which it is "a target cell". The control of this selective gene expression is understood little, but it is immensely important. As an absurd example of what might happen if it went wrong, consider the endocrine mayhem that would ensue if the various cell populations of the pituitary that secrete hormones were randomly to interchange their expressed receptors and thus secrete their hormones in response to the wrong hypothalamic releasing factors. Moreover, rather than being static, each cell's receptor status is constantly susceptible to environmental regulation. For example, when antigens stimulate lymphocytes to proliferate and then differentiate into mature immune effector cells this complicated sequence of events depends on the correct sequential expression in the developing cells of a series of receptors for lymphokines.

The onward transmission of receptor signals to the cell interior: general principles

The considerable genetic load on the cell of encoding a huge variety of receptors would be further exacerbated if each type of activated receptor had its own unique machinery for transmitting its, and only its, message onwards into the cell. Fortunately, as summarised in figure 1, this is not the case. Instead, the biochemical machinery of the plasma membrane translates the immensely complex language of extracellular stimuli to which all cells are exposed into a very much simpler language of a limited number of ionic, electrical, and chemical signals that control intracellular events. Some of the better understood mechanisms, and some of the receptors that use them, are summarised in figure 2 (a more detailed version of which has been published elsewhere[1]), and table II lists some of the receptors that use these and other fairly well characterised signalling mechanisms. These signalling mechanisms have been very well described in a recent book,[3] and our rapidly advancing knowledge of these mechanisms is updated annually in a suite of brief reviews in *Current Opinion in Cell Biology*.[4]

As shown in figure 2, some receptor proteins, which probably constitute a minority of the cell's total receptor repertoire, incorporate outward facing recognition sites (through which they respond to extracellular stimuli) and also the mechanisms needed for the onward transmission of the stimulus to the cell interior. Most of the receptors of this general design that have so far been identified fall into one of two major receptor families, the ligand-gated ion channels or receptor tyrosine kinases (see below). The ligand-stimulated guanylate cyclases form a third, smaller, group (table II).[5,6]

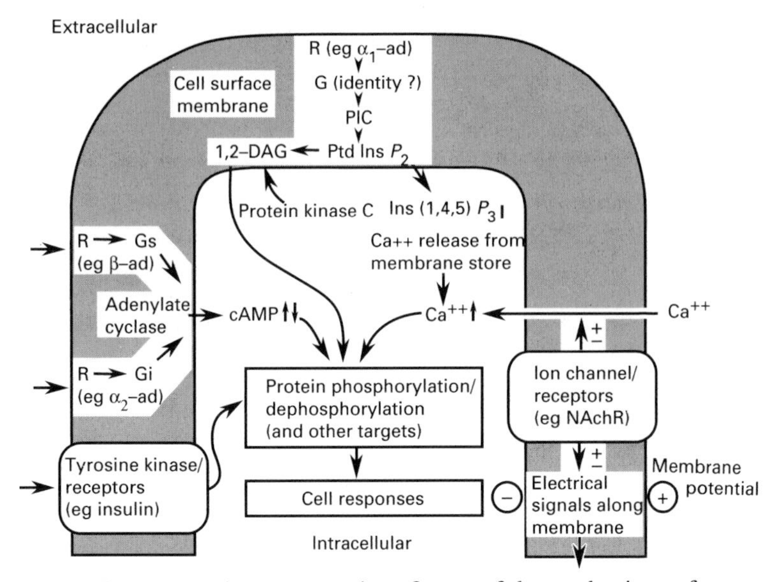

FIG 2—Diagrammatic representation of some of the mechanisms of transmembrane signalling used by cell surface receptors. At the left are shown receptors possessing intrinsic protein tyrosine kinase activity, alongside the control of adenylate cyclase by stimulatory and inhibitory receptors (indicated as R) that is mediated by G_s and G_i. At the top is the receptor-stimulated hydrolysis of PtdIns (4,5)P_2 (phosphatidylinositol 4,5-bisphosphate) by phosphoinositidase C (PIC) leading to intracellular accumulation of 1,2-diacylglycerol (1,2-DAG), Ins(1,4,5)P_3, and Ca^{++}. At the right are the actions of receptors possessing intrinsic ion channels on membrane potential and hence on cellular electrical activity and Ca^{++} permeability. Many, but not all, of the intracellular effects of changes in the cellular concentrations of cyclic AMP, 1,2-DAG, and Ca^{++} are caused by the phosphorylation and dephosphorylation of particular target proteins by the various kinases and phosphatases that these intracellular messengers regulate, as depicted in the centre.

However, many more cell surface receptors signal through relay systems in which three components—namely the receptor, a coupling protein that depends on guanine nucleotide (G protein), and an effector protein, which is either an enzyme or an ion channel—act in sequence (see figs 1 and 2, table II). The receptors that use this type of mechanism all appear to be members of a single structural family of polypeptides. They are embedded within the plasma membrane, which their polypeptide chains traverse seven times. The ligand recognition sites of these "7-span" receptors are exposed to the extracellular medium, and their coupling to G proteins (and thence to effectors) is achieved primarily through the

TABLE II—*Some examples of the mechanisms of signal transmission used by receptors*

Receptors with intrinsic ion channels[11-14]	
Excitatory: Acetylcholine (nicotinic receptors with Na^+ channels), glutamate (some receptors)	
Inhibitory: γ-aminobutyric acid (GABA) and glycine (with Cl^- channels)	
Receptors with intrinsic protein tyrosine kinase activity[15-20]	
Platelet derived growth factor (PDGF)	Insulin
Macrophage colony stimulating factor (M-CSF)	Epidermal growth factor (EGF)
	Insulin-like growth factor 1 (IGF-1)
Haemopoietic stem cell factor (SCF/*Steel* factor)	Hepatocyte growth factor (HGF)
	Nerve growth factor (NGF)
Receptors with intrinsic guanylate cyclase activity[5,6]	
Atrial natriuretic peptide	*E coli* enterotoxin receptor (gut)
G protein-coupled receptors that activate adenylate cyclase[6-9]	
Adrenergic (β_1 and β_2 receptors)	Adrenocorticotrophic hormone (ACTH)
Vasopressin (V_2 receptor)	
Glucagon	Prostacyclin
Thyroid stimulating hormone (TSH)	Growth hormone releasing hormone (GHRH)
Receptors mediating olfaction (a large family)	Some taste receptors?
Histamine (H_2 receptor)	
G protein-coupled receptors that inhibit adenylate cyclase[6-9]	
Adrenergic (α_2 receptor)	Opioid peptides (δ receptor)
Acetycholine (muscarinic–two subtypes)	Prostaglandin E_1 (some tissues)
	Adenosine (A_1 receptor)
G protein-coupled receptors that activate a cyclic nucleotide phosphodiesterase[2,5,6,8,9]	
Rhodopsin (vertebrates: four forms, in retinal rods and three types of colour-selective cones)	
Some taste receptors?	
G protein-coupled receptor that opens a hyperpolarising K^+ channel[8,9]	
Acetylcholine (muscarinic—one subtype)	
G protein-coupled receptors that activate PtdIns(4,5)P_2 hydrolysis[7-9,21-28]	
Adrenergic (α_1 receptor)	Glutamate ("metabotropic" receptor)
Acetylcholine (muscarinic–two subtypes)	Histamine (H_1 receptor)
	Vasopressin (V_1 receptor)
Substance P/other tachykinins	Endothelins (three receptors)
Thyrotropin releasing hormone (TRH)	Thromboxane A_2
Platelet activating factor (PAF or AGEPC)	Rhodopsin (invertebrates)
	Some taste receptors?

intracellular polypeptide loop between the fifth and sixth transmembrane domains.[7] The various G proteins to which these receptors pass their information constitute another, but smaller, evolutionary family, and the effector proteins that are activated by the G proteins are much more structurally and functionally diverse (table II).[89] Two major evolutionary achievements of these receptor and G protein superfamilies are: 1) that they allow an enormously diverse range of extracellular signals to channel cellular control information through a limited number of effector molecules; and 2) that they facilitate the evolutionary emergence of multiple receptors for single stimuli, often using different signalling mechanisms (for example, the various muscarinic acetylcholine receptors, see table II).

Signalling through the opening of cell surface ligand-gated ion channels

The main permeability barrier within cell surface membranes is provided by the hydrocarbon side chains of membrane lipids, and the membranes normally have a very low permeability to small ions such as Na^+, Ca^{++}, and Cl^-. Membrane pumps driven by ATP constantly build ion gradients across this barrier, notably to keep the intracellular Na^+ concentration at least 10-fold lower than the 100 mmol/l of extracellular fluid and the intracellular Ca^{++} concentration 10 000-fold lower than the extracellular Ca^{++} concentration of 1–2 mmol/l. These ion gradients, and others dependent on them such as the excess of extracellular over intracellular Cl^-, permit two types of signalling controlled by receptors. Firstly, they give rise to a resting membrane potential (with the intracellular surface of the membrane electrically negative relative to the outside) that is perturbed whenever there are selective and transient increases in membrane permeability to particular ions. Secondly, the Ca^{++} concentration in the cytoplasm is held so low (at about 0·1 μmol/l) that small ion movements can raise this value substantially: a relatively small rise in cytoplasmic Ca^{++} can therefore serve as an effective intracellular signal.

The prototype of the ligand-gated ion channel receptors, which are multisubunit proteins incorporating transmembrane ion channels that open transiently when the receptors bind an agonist, is the fast acting nicotinic acetylcholine receptor (NAChR) of skeletal muscles and ganglia. This was one of the first receptors to be obtained in sufficient quantity for detailed biochemical and structural study, because it is strikingly abundant in the electric organs of electric eels and rays (these organs are evolutionary simplifications of skeletal muscles, retaining little other than the receptor rich neuromuscular junctions of the progenitor muscles). The receptor is a five subunit ($\alpha_2\beta\gamma\delta$) protein, with a molecular weight of about 280 000, that spans the cell surface membrane from outside to inside. Multiple isoforms of its individual polypeptide chains are encoded by different genes, so allowing different cells (fetal and adult muscles, neurones of various types, etc) to express subtly different types of nicotinic acetylcholine receptor.[10 11] Each subunit of the receptor spans the membrane and contributes amino acid residues to a protected pathway that can pass cations, particularly Na^+, rapidly through the membrane barrier. When acetylcholine binds to the extracellular portions of the α subunits the ion channel opens for a few milliseconds, allowing Na^+ to flow down its gradient and thus reduce the membrane potential. The best known result of such membrane depolarisation is the initiation of long distance electrical signals in the form of action potentials, as in nerve cells and at the neuromuscular junction. A major effect in some other cells, however, is to cause local changes in membrane permeability to key ions. For example, stimulation of the nicotinic acetylcholine receptor causes catecholamine secretion from the adrenal medullary cells of some species, because the membrane depolarisation triggered by receptors opens plasma membrane Ca^{++} channels that are controlled by membrane potential and thus leads to a Ca^{++} triggered exocytosis of the hormone. A second set of structurally similar receptors, including the so called NMDA (N-methyl-D-aspartate) receptors, mediate neuronal responses to excitatory amino acids such as glutamate. These receptors are involved both in early stages of memory formation and in the neurodegenerative processes that are triggered if neurones are persistently excited to an excessive degree.[12–14]

The receptors for the inhibitory neurotransmitters γ-aminobutyric acid (GABA) and glycine also include an ion channel: these receptors belong to the same evolutionary superfamily as the excitatory receptors considered in the previous paragraph.[10 11] However, when these channels open in response to neurotransmitters the negatively charged ion, Cl^-, flows into the "stimulated" cells. Because this Cl^- flux tends to raise the membrane potential it increases the resistance of neurones to activation.

Receptors with intrinsic or associated protein tyrosine kinase activity

The clearest examples of receptor proteins that exhibit an intrinsic enzyme activity are the receptor tyrosine kinases

(RTKs).[15 16] The first known members of this family, the platelet derived growth factor (PDGF) and epidermal growth factor (EGF) receptors (see table II), were isolated as receptor proteins that catalysed the tyrosine phosphorylation both of their own cytoplasmic domains and of other proteins. Amino acid sequencing of the epidermal growth factor receptor showed that the previously cloned v-erbB oncogene encodes a constitutively active derivative of this receptor. These discoveries were quickly followed by the identification—sometimes through the direct cloning of oncogenes and sometimes through homology cloning using oligonucleotides representing consensus elements in the first sequences determined—of many receptor tyrosine kinases for which the natural ligands were unknown. Some of these "orphan receptors" have since had their functions defined: for example, the met, trk, fms, and kit proto-oncogenes encode, respectively, receptors for hepatocyte growth factor (HGF), nerve growth factor (NGF), macrophage colony stimulating factor (M-CSF), and stem cell factor (SCF).[17–19]

All of these kinases incorporate both an extracellular ligand binding site (for example, for a growth factor) and a kinase domain at the cytoplasmic surface of the plasma membrane. In most receptor tyrosine kinases, such as the epidermal growth factor receptor, these two functional elements are towards the opposite ends of a single polypeptide chain that spans the membrane, but the insulin and insulin-like growth factor 1 (IGF-1) receptors exist as disulphide bridged $\alpha_2\beta_2$ tetrameric proteins that include two ligand binding and two kinase sites. Stimulation of receptor tyrosine kinases by their cognate ligands activates their tyrosine kinase activity, causing receptor phosphorylation on the cytoplasmic domain and then phosphorylation of other protein substrates. The initial phosphorylation of the receptors is achieved by interchain transphosphorylation, in which one chain of a receptor dimer phosphorylates the other: in the insulin and insulin-like growth factor 1 receptors this occurs within the pre-existing receptor oligomer, whereas dimeric interactions of other receptors are driven by ligand binding. Once the receptor is phosphorylated, its kinase site becomes opened up to allow phosphorylation of other proteins.

Although it is now known that the tyrosine kinase activity of receptor tyrosine kinases is essential for their biological effects, it is still not clear how expression of the tyrosine kinase activity controls complex cell behaviours such as nuclear gene activation.

What is clear is that many receptor tyrosine kinases stimulated by growth factors do phosphorylate and physically interact with a suite of key proteins that are somehow implicated in growth control: these include GAP (a guanosine triphosphatase (GTPase) activating protein that is a functional partner of the $p21^{ras}$ product of c-ras proto-oncogenes), the regulatory subunit of a 3-kinase which converts PtdIns(4,5)P_2 (phosphatidylinositol 4,5-bisphosphate) to PtdIns(3,4,5)P_3, the γ_1 isozyme of phosphoinositidase C (PIC), and tyrosine kinases of the src family.[16] Most of these proteins share sequence motifs that are known as SH2 (SH = src homology) domains, which interact direct with sequences that contain phosphotyrosine in the activated receptor tyrosine kinases. Downstream of these events, there is activation of a family of serine or tyrosine kinases (known as either mitogen activated protein (MAP) kinases or extracellular signal-regulated kinases (ERKs)) which seem to be essential, in some as yet undefined manner, for cell responsiveness.[20]

The transmembrane receptor tyrosine kinases are not the only cell surface receptors for whose function the activation of a tyrosine kinase is essential. Some other receptors, of which the multisubunit $\alpha\beta$/CD3($\gamma\varepsilon\zeta\eta$) antigen receptor complex of T lymphocytes is the best understood, are non-covalently complexed to and activate tyrosine kinase molecules of the src family, through which they transmit their activation message in a manner analogous to the action of the receptor tyrosine kinases.[21]

Receptors that control the formation of intracellular second messenger molecules

About 35 years ago Earl Sutherland and his colleagues discovered that catecholamines (acting through β receptors), glucagon, and various other hormones (see table II) transmit their messages to the cell interior by stimulating the synthesis by adenylate cyclase of cyclic AMP, an intracellular nucleotide "second messenger" that is present at only micromolar concentrations. In its turn, a rise in the intracellular concentration of cyclic AMP activates a protein kinase that phosphorylates (and thus activates or inactivates) key enzymes of hormone-regulated metabolic pathways, including those responsible for muscle glycogen breakdown and adipose lipolysis. At the time of this discovery Sutherland acknowledged that cyclic AMP was likely to be only the first of many intracellular "second messengers" that mediate hormone actions, but he could hardly have guessed how long it would take to identify others or the extent to which work on adenylate cyclase would come to influence the studies of other signalling systems.

A substantial proportion of the many 7-span receptor species at the cell surface are now known to control cells by influencing adenylate cyclase activity: some stimulate cyclic AMP formation whereas others inhibit it (table II).[7-9] Thus all of the extracellular information delivered to these diverse receptors is summarised for the cell interior simply as a rise or fall in the intracellular concentration of cyclic AMP. The first stage of this simplification is a division of receptors into a stimulatory group and an inhibitory group, each of which communicates with adenylate cyclase through a different coupling protein that is dependent on guanosine triphosphate (GTP): these stimulatory and inhibitory G proteins are known respectively as G_s and G_i. Although it is clear that activation of the $\alpha\beta\gamma$ heterotrimeric G proteins by stimulated receptors provokes their dissociation into free α subunits and $\beta\gamma$ complexes, with the quantities of the free α_s and α_i within the membrane regulating adenylate cyclase activity, the molecular details of the stimulatory and inhibitory interplay between G_s, G_i, and adenylate cyclase remain a matter of some argument.

A second large family of 7-span receptors feeds its input to the cell through a very different signalling system based on the hydrolysis of the minor membrane phospholipid component, PtdIns(4,5)P_2, by phosphoinositidase C (fig 2). As long ago as 1953, Hokin and Hokin discovered that stimulation of receptors activates inositol phospholipid metabolism, and I recognised in 1975 that inositol lipid hydrolysis somehow leads to a rise in cytoplasmic Ca^{++} concentration,[22] but it was not until the early 1980s that the exact role of PtdIns(4,5)P_2 hydrolysis in transmembrane signalling was defined.[23 24]

As with adenylate cyclase, control of phosphoinositidase C activity by most receptors is indirect, with G proteins serving as intermediaries. The coupling of some receptors (for example, that for the neutrophil chemotactic peptide, *f*methionine, leucine, phenylalanine (*f*MetLeuPhe)) to this enzyme is abolished by treating cells with pertussis toxin, which prevents certain G proteins from responding to "their" receptors, whereas receptor activation of phosphoinositidase C by other receptors (for example, the V_1 vasopressin receptor) survives this treatment. It therefore seems likely that at least two G proteins can couple receptors to the enzyme's activation.[9] The G protein resistant to pertussis toxin was recently identified as G_Q (also known as G_{11}), which activates the β isozyme of phosphoinositidase C, but the G protein sensitive to pertussis toxin and the phosphoinositidase C isozyme that it activates remain to be defined.

Hydrolysis of PtdIns(4,5)P_2 catalysed by phosphoinositidase C yields two products: water soluble inositol 1,4,5-triphosphate (Ins(1,4,5)P_3), which diffuses into the cell interior; and lipid soluble 1,2-diacylglycerol (1,2-DAG), which remains associated with the cytoplasmic surface of the cell membrane.[23–25] Each of these products has a unique role as an intracellular messenger. Within cells there is a membrane compartment, probably a portion of the endoplasmic reticulum, into which Ca^{++} is continuously pumped by an ATP driven pump, so holding the cytoplasmic Ca^{++} concentration of "resting" cells at about 0·1 μmol/l. When receptors cause a rise in the intracellular concentration of Ins(1,4,5)P_3 this compound binds to its own receptors on the membrane enclosing this Ca^{++} store and triggers a rapid release of Ca^{++} into the cytoplasm. The receptor proteins that respond to Ins(1,4,5)P_3 are a small family of ligand-gated Ca^{++} channels consisting of oligomers (possibly tetramers) of a 250 000 Da polypeptide: they probably function in a manner basically similar to the neurotransmitter-regulated ion channels of the plasma membrane.[26–28] As a result of stimulation, the cytoplasmic Ca^{++} concentration often rises briefly to $>0{\cdot}5$ μmol/l within seconds after the application of a stimulus. Ins(1,4,5)P_3 is inactivated both by dephosphorylation and by entry into a complex series of pathways that interconvert several previously unknown inositol polyphosphate isomers, the biological functions of which are still uncertain.[23 25] Downstream of the mobilisation of Ca^{++} stimulated by Ins(1,4,5)P_3, cells often show temporally and spatially complex changes in cytosolic Ca^{++} concentration, most notably temporally separated spikes of $[Ca^{++}]$ whose frequency is regulated by the agonist concentration and travelling waves of $[Ca^{++}]$ that traverse the cell from an initiating site that is thought to be a Ca^{++} store responsive to Ins(1,4,5)P_3.[23 26 27] 1,2-Diacylglycerol, the other messenger molecule formed from PtdIns(4,5)P_2, activates one or more of a family of isozymic protein kinases (the protein kinases C; PKC) that can also be activated by phorbol ester tumour promoters. The latter compounds were originally of interest because they enhanced the tumour yield in skin treated with carcinogens, but they were later found partially to mimic many effects of cell stimulation (such as platelet aggregation and lymphocyte proliferation), suggesting that they might act by subverting some normal cellular signalling process.[24]

Some of the intracellular targets of Ca^{++} and protein kinase C are known—for example, Ca^{++} (in combination with the intracellular Ca^{++} receptor protein, calmodulin) activates glycogen breakdown, smooth muscle contraction, and the pumping of excess Ca^{++} from cells; protein kinase C controls a plasma membrane ion channel that exchanges intracellular K^+ ions for extracellular Na^+ and thus controls intracellular pH. However, we do not yet fully understand either the mechanisms by which these two signals often synergise in the activation of cells or those by which they control longer term cellular responses.

The future

Past research has identified, and given us a partial understanding of, some of the biochemical mechanisms by which cells respond to extracellular controls. Several are discussed briefly above, other well established mechanisms and some receptors that use them are mentioned in table II, and yet others are either only just emerging or are yet to be discovered. In the future we must continue to explore the details of the known transmembrane signalling processes, identify an as yet unknown number of others, and in particular determine how the intracellular messengers they generate control complex cell behaviours such as the selective gene expression that is essential to successful cell differentiation. There is evidence, including the recent identification by cloning of a substantial number of "orphan" G proteins[7] that await assignment of functions, that yet more signalling reactions will turn out to be under G protein control: candidates include a phospholipase A_2 (which liberates free arachidonate and thus initiates eicosanoid synthesis);[29] and additional C type phospholipases that liberate 1,2-diacylglycerol and ceramide (N-acylsphingosine), respectively, from phosphatidylcholine and sphingomyelin—1,2-diacylglycerol might lead to activation of protein kinase C without an accompanying rise in intracellular Ca^{++} concentration,[30] and the ceramide formed from sphingomyelin may have its own messenger role.[31] Doubtless other mechanisms will emerge, some of them from quite unexpected quarters, over the next few years.

1 Michell RH. Centrefold on transmembrane signalling. *Trends in Pharmacological Sciences* 1988, April issue.
2 McLaughlin SK, McKinnon PJ, Margoskee RF. Gustducin is a taste-cell-specific G protein closely related to the transducins. *Nature* 1991;**357**:563–9.
3 Barritt GJ. *Communication within animal cells*. Oxford: Oxford University Press, 1992.
4 Annual review issues on cell regulation. *Curr Opin Cell Biol* 1989;**1**:157–235, 1990;**2**:165–237, 1991;**3**:169–234, 1992;**4**:141–273.
5 Thompson DK, Garbers DL. Guanylyl cyclase in cell signalling. *Curr Opin Cell Biol* 1990;**2**:206–11.
6 Bentley JK, Beavo JA. Regulation and function of cyclic nucleotides. *Curr Opin Cell Biol* 1992;**4**:233–40.
7 Iismaa TP, Shine J. G protein-coupled receptors. *Curr Opin Cell Biol* 1992;**4**:195–202.
8 Bourne HR, Sanders DA, McCormick F. The GTPase superfamily: conserved structure and molecular mechanism. *Nature* 1991;**349**:117–27.
9 Spiegel AM. G proteins in cellular control. *Curr Opin Cell Biol* 1992;**4**:203–11.
10 Miller C. Genetic manipulation of ion channels: a new approach to structure and mechanism. *Neuron* 1989;**2**:1195–205.
11 Unwin N. The nature of ion channels in membranes of excitable cells. *Neuron* 1989;**3**:665–76.
12 Mayer M. Fine focus on glutamate receptors. *Current Biology* 1992;**2**:23–5.
13 Collingridge GL, Singer W. Excitatory aminoacid receptors and synaptic plasticity. *Trends Pharmacol Sci* 1990;**11**:290–7.
14 Bliss TVP, Collingridge GL. A synaptic model of memory: long-term potentiation in the hippocampus. *Nature* (in press).
15 Waterfield MD. Growth factor receptors. *Br Med Bull* 1989;**45**:541–53.
16 Cantley LC, Auger KR, Carpenter C, Duckworth B, Graziani A, Kapeller R, *et al*. Oncogenes and signal transduction. *Cell* 1991;**64**:281–302.
17 Park M. Lonesome receptors find their mates. *Current Biology* 1991;**1**:248–50.
18 Wagner EF, Alexander WS. Of Kit and mouse and man. *Current Biology* 1991;**1**:356–8.
19 Gherardi, E, Stoker M. Hepatocyte growth factor/scatter factor: mitogen, motogen and *met*. *Cancer Cells* 1991;**3**:227–32.
20 Maller J. MAP kinase activation. *Curr Biol* 1991;**1**:334–5.
21 Klausner RD, Samelson LR. T antigen receptor activation pathways: the tyrosine kinase connection. *Cell* 1991;**64**:875–8.
22 Michell RH. Inositol phospholipids and cell surface receptor function. *Biochim Biophys Acta* 1976;**415**:81–147.
23 Berridge MJ, Irvine RF. Inositol phosphates and cell signalling. *Nature* 1989;**341**:197–205.
24 Nishizuka Y. The molecular heterogenity of protein kinase C and its implications for cellular regulation. *Nature* 1988;**334**:661–5.
25 Downes CP, Macphee CH. Myo-inositol metabolites as cellular signals. *Eur J Biochem* 1990;**193**:1–18.
26 Tsien RW, Tsien RY. Calcium channels, stores and oscillations. *Annu Rev Cell Biol* 1990;**6**:715–60.
27 Lytton J, Nigam SK. Intracellular calcium: molecules and pools. *Curr Opin Cell Biol* 1992;**4**:220–6.
28 Mikoshiba K, Furuichi T, Miyawaki A, Yoshikawa S, Maeda N, Niinobe M, *et al*. The inositol 1,4,5-trisphosphate receptor. *Ciba Found Symp* 1992;**164**: 17–35.
29 Axelrod J, Burch RM, Jelsema CL. Receptor-mediated activation of phospholipase A_2 via GTP-binding proteins: arachidonic acid and its metabolites as second messengers. *Trends Neurosci* 1988;**1**:117–23.

30 Billah MM, Anthes JC. The regulation and cellular function of phosphatidylcholine hydrolysis. *Biochem J* 1990;**269**:281–991.
31 Kolesnick R. Ceramide, a novel second messenger. *Trends Biochem Sci* (in press).

Cell to cell and cell to matrix adhesion

D R Garrod

The structure and organisation of body tissues is dependent on maintenance of contact between cells and their neighbours and between cells and the extracellular matrix. In simple epithelia, such as the lining of the intestine or the kidney tubule, the individual cells have surfaces with three different sets of adhesive properties. The apical or luminal surface is non-adhesive, the lateral surface is specialised for adhesion to adjacent cells, and the basal surface is specialised for adhesion to the underlying matrix, the basement membrane.

In a stratified epithelium, such as epidermis, the cells in the basal layer adhere to the basement membrane below, to each other laterally, and to suprabasal cells apically. The suprabasal cells have lost adhesion to the matrix and instead adhere to similar cells on all sides. Apically this mutual adhesion is also lost and cells are sloughed off.

Other cell types—leucocytes and blood platelets—spend much of their time circulating freely and thus showing no adhesive interactions. Lymphocytes, however, show quite specific recirculation patterns in which particular subsets leave the blood circulation by first adhering to high endothelial venule cells at specific sites—for example, peripheral lymph nodes or mucosal associated lymphoid tissue. They then migrate into the lymphoid tissue and eventually return to the blood circulation. Endothelial adhesion by leucocytes is also the first step in a range of adhesive interactions required in their tissue invasive response to inflammation. Blood platelets respond to injury by a whole set of adhesive interactions with endothelial cells, with the matrix of the clot and endothelial basement membrane, and with each other.

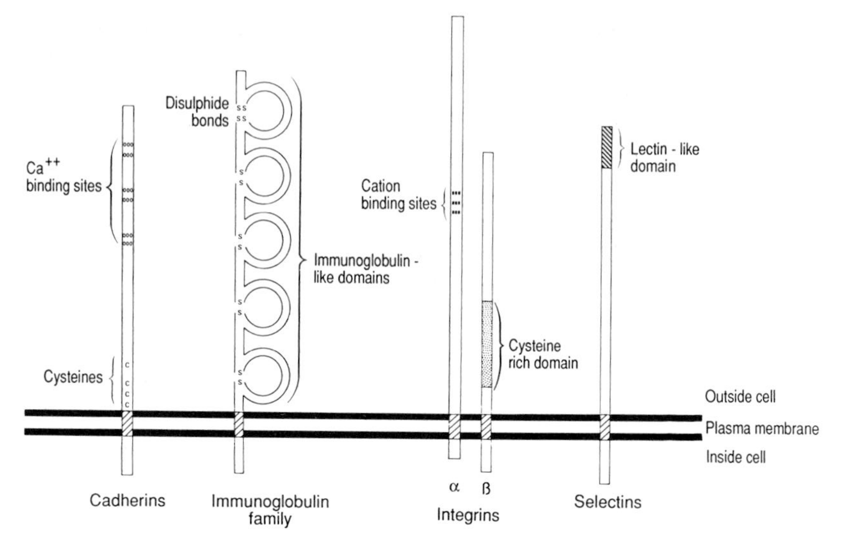

Molecular structure of families of cell adhesion molecules

Thus cells of any particular type possess a set of adhesive properties that may be both spatially and temporally regulated. A considerable amount is now known about the molecular mechanisms that mediate these properties.[1 2]

Families of cell adhesion molecules

Many different cell adhesion molecules have been described. Most of them belong to one of a small number of families of related molecules in which the individual members share the same basic molecular structure but are subtly different from each other (figure).

Cadherins

In most tissues a major contribution to cell to cell adhesion is made by calcium dependent cell adhesion molecules known as cadherins.[3] In general these are simple transmembrane glycoproteins. The extracellular domain has an adhesion site towards the N-terminal region and several calcium binding sites. Adhesive binding is homophilic: a cadherin molecule on one cell binds to another cadherin molecule of the same type on the next cell. Linkage of the cytoplasmic domain to the cytoskeleton through proteins known as catenins is necessary for cadherin function. The best characterised is epithelial cadherin, E-cadherin, or uvomorulin. This appears very early in development, when it is involved in compaction of the eight cell embryo and cell polarisation. In adult epithelia—for example, intestinal epithelium, it is present on the lateral cell surfaces but is concentrated in intercellular junctions known as the zonulae adherentes, which ring the cells in the apicolateral region. The zonula adherens is characterised by a cortical ring of cytoskeleton, the major component of which is actin.

The adhesive glycoproteins of the other major intercellular junctions of epithelia, the desmosomes, are members of the cadherin family. Their extracellular domains are very like those of cadherin but their cytoplasmic domains differ, being specialised for forming desmosomal plaques and, thereby, attachment to the keratin intermediate filament cytoskeleton rather than to actin.

A cell adhesion molecule known as neural cadherin (N-cadherin) is expressed predominantly in the nervous system, where it is implicated in neuronal migration during development.

Immunoglobulin superfamily

The other major group of cell to cell adhesion molecules are members of the immunoglobulin superfamily.[4] The extracellular portions are characterised by the presence of at least one, and usually multiple, immunoglobulin-like domains. Included in this group are several nervous system adhesion molecules such as the neural cell adhesion molecule (N-CAM), L1, and TAG, which are involved in neuronal guidance and fasciculation. Several members of the immunoglobulin family are concerned with antigen recognition and adhesion in T lymphocytes. These include the T cell receptor (CD3) and its coreceptors CD4 and CD8, which together recognise complexes of antigen peptide and major histocompatibility complex on other cells; the major histocompatibility complex molecules themselves; and lymphocyte function-related antigen 2 (LFA-2 or CD2), a receptor for another immunoglobulin-like molecule, LFA3, expressed on other cells. Another group of immunoglobulin-like cell adhesion molecules includes the so called intercellular adhesion molecules, ICAM-1 and ICAM-2, which are more widely expressed, for example, on epithelial and endothelial cells, and V-CAM (which is expressed on endothelial cells). These three are involved in the inflammatory response.

The immunoglobulin superfamily is large and diverse probably because the basic structure of the immunoglobulin domain is versatile and readily adaptable to different binding functions. Among these molecules, however, only the T cell receptor and the immunoglobulins themselves have somatically variable domains necessary for antigen recognition. Members of the superfamily are present in insects, where they are involved in forming nerve connections; thus the association of immunoglobulin-like domains in cellular recognition preceded the immune system in evolution.

Integrins

Both cell to cell and cell to matrix receptors are contained within the remaining large family of adhesion molecules, the integrins. These are heterodimers consisting of one α chain and one β chain, both of which are necessary for adhesive binding. Fourteen different α chains and eight different β chains are now known. Integrins may be classified into subfamilies according to which β subunit is involved in the complex. Thus $\beta1$ integrin may associate with one of eight different α subunits to give a series of matrix receptors of differing specificity. The $\beta2$ integrins on the other hand are a family of cell to cell adhesion molecules of lymphoid cells with three alternative α subunits. The classification is made more complicated because some α subunits can associate with different β subunits (for example, $\alpha6\beta1$ and $\alpha6\beta4$).

Some integrins are apparently quite specific in their ligand-binding properties—for example, $\alpha5\beta1$ for the arginine, glycine, and aspartic acid (-arg-gly-asp-) tripeptide sequence of fibronectin—whereas others are promiscuous—for example, $\alpha v\beta3$, once regarded as the vitronectin receptor, also binds fibronectin, fibrinogen, von Willebrand factor, thrombospondin, and osteopontin. An interesting example is $\alpha4\beta1$, which binds both the IIICS domain of fibronectin and the immunoglobulin-like molecule V-CAM on endothelial cells. To complicate matters further individual cell types usually express multiple integrins. A good example to consider here is the blood platelet that expresses predominantly αIIb$\beta3$ (GPIIb/IIIa), which binds fibrinogen, fibronectin, von Willebrand factor, and vitronectin but also lesser amounts of $\alpha V\beta 3$, $\alpha 5\beta 1$, $\alpha 2\beta 1$ (collagen), and $\alpha 6\beta 1$ (laminin).

Selectins

Most cellular adhesive interactions seem to entail homophilic or heterophilic protein to protein binding. However, the final, as yet small, family of cell adhesion molecules bind to carbohydrate. These are the selectins, which have lectin-like domains at their extracellular N-terminal extremities. One of these, L-selectin (previously LAM-1/Mel-14), is a "homing receptor," mediating specific adhesion of lymphocytes to endothelium in peripheral lymph nodes. This molecule is also involved in the adhesion of neutrophils to endothelium during the inflammatory response. Two other members of this family, E-selectin (previously ELAM-1, endothelial leucocyte adhesion molecule) and P-selectin (previously GMP-140, PADGEM or CD62), also participate in the inflammatory response. E-selectin is up regulated on endothelial cells over a period of hours after stimulation by inflammatory mediators. P-selectin is contained within Wiebel-Palade bodies of endothelial cells and platelet α granules, from which it is rapidly mobilised on activation or clotting, mediating adhesion to neutrophils and monocytes.

Cell adhesion and disease

Leucocyte adhesion deficiency

Several rare diseases result from specific defects in adhesion molecules. The inherited immunodeficiency disease leucocyte adhesion deficiency is characterised by pronounced granulocytosis, lack of pus formation, mobilisation of neutrophils and monocytes to inflammatory sites, severe gingivitis, and recurrent or progressive soft tissue infections. It is caused by defects in the common $\beta 2$ subunit of the lymphocyte integrins, resulting for example, in deficient expression of the $\alpha\beta$ heterodimers at the cell surface with the result that adhesion of lymphocytes is impaired. Defects in $\beta 2$ may be either mutations, which affect the structure of the protein, or greatly reduced expression or absence of messenger RNA (mRNA). In its most severe form the disease causes death in early childhood from overwhelming infection. It has been shown recently that adhesiveness of lymphocytes from patients with leucocyte adhesion deficiency can be restored by transfection, with an expression vector continuing the complementary DNA for the normal β2 subunit. Thus gene therapy may be feasible for these patients.

Mutation of the β3 subunit of the platelet integrin GPIIb/IIIa results in the bleeding disorder Glanzmann's thrombasthenia. The Bernard-Soulier syndrome entails a deficiency of the non-integrin platelet adhesion receptor GPIb/IX and von Willebrand's disease a deficiency of the receptor WF.

Pemphigus

In pemphigus, the group of autoimmune diseases that cause epidermal blistering, patients' serum contains autoantibodies to desmosomes, the cadherin-like glycoproteins of the epidermal intercellular junctions. The autoantibodies cause the junctions to break down and the keratinocytes to separate (acantholysis) giving rise to the formation of blisters. Breakdown of desmosomes by another mechanism (possibly protease activity) gives rise to the inherited Darier's disease and Hailey-Hailey disease, which are similar but probably unrelated.

Failure of adhesion between the epidermis and its underlying basement membrane occurs in bullous pemphigoid, in which autoantibodies to cytoplasmic protein compounds of the matrix adhesive junctions known as hemidesmosomes are found in serum.[5] The hemidesmosomes adhesion receptor is another integrin, α6β4. The matrix ligand for this may be a newly discovered trimeric basement membrane component known as BM-600, nicein, kalinin, or epiligrin. (Like many new proteins this has been discovered by different groups, which have given it different names.) Another group of blistering diseases, epidermolysis bullosa, results from failure of epidermis to adhere to basement membrane. The most severe, junctional epidermolysis bullosa, is fatal in early childhood. It entails defects in hemidesmosomes, possibly resulting from defective interaction between the integrin and its ligand.

Metastatic behaviour

Defective cell adhesion has long been thought to play a part in the invasive and metastatic behaviour of neoplastic cells. Invasive cells spread into the tissues surrounding the primary tumour; penetrate into blood vessels, lymphatics, or body cavities; and become dispersed to distant areas. Some may become trapped at new sites, extravasate, and form secondary tumours. This is a complex series of events that may entail various altered cellular properties, such as secretion of proteolytic enzymes, alterations in cell motility, and altered growth properties, as well as possible changes in adhesiveness. However, some interesting observations have been made in relation to adhesion and tumour spread. For example, formation of secondary tumours by injected melanoma cells in mice was inhibited by simultaneous injection of a short synthetic peptide containing the -arg-gly-asp- tripeptide that blocks integrin $\alpha 5\beta 1$ binding to fibronectin,[6] strongly suggesting a role for matrix adhesion in the formation of metastases. A series of small cysteine-rich peptides that contain -arg-gly-asp- (which are called disintegrins) has been identified in viper venom.[7] These peptides are much more potent inhibitors of $\beta 1$ and $\beta 3$ integrins than synthetic peptides and inhibit tumour metastasis in mice, as well as platelet aggregation.

In a series of human tumour cell lines invasion into collagen gels was inversely related to expression of E-cadherin and inhibition of invasion was achieved in a breast carcinoma cell line by transfection with a vector expressing E-cadherin.[8] This suggests a possible tumour suppressor function for this cell adhesion molecule. In a detailed study of the genetic deletions found in colorectal cancer one of the commonly deleted genes (DCC) was found to code for a protein homologous to the immunoglobulin superfamily of cell adhesion molecules, again suggesting a possible tumour suppressor function.[9]

Viruses and parasitic diseases

Cell adhesion molecules also participate in some infectious and parasitic diseases, in which the invading organisms use the normal tissue molecules for binding. Thus the rhinoviruses—RNA viruses responsible for about a half of common colds—bind to intercellular adhesion molecules on respiratory epithelium. Of more sinister importance, the binding of human immunodeficiency virus (HIV) during infection of T lymphocytes is mediated by binding of the viral gp120 protein to CD4, the immunoglobulin-like T cell coreceptor. Intercellular adhesive molecules are also implicated in the adhesion of red blood cells infected with *Plasmodium falciparum* to capillary endothelium in the pathogenesis of severe malaria.[10] Cytokine mediated up regulation of endothelial intercellular adhesion molecules may be associated with severity of the disease.

Other important functions

An important function of cell adhesion molecules in addition to adhesive binding is transmission of signals across cell membranes. Study of this has so far been limited, but several aspects have emerged. Members of all three of the large families of cell adhesion molecules—the cadherins, integrins, and immunoglobulin-like superfamily—have been shown to participate in signalling. Signals can be transmitted from the outside to the inside of cells in response to ligand binding, affecting second messengers and gene transcription, or from the inside to the outside of cells, modulating the binding affinity of cell adhesion molecules. Add to this the quantitative regulation of expression of cell adhesion molecules in response, for example, to inflammatory mediators and associated with changes in cellular differentiation, and we find a complex and highly responsive set of cellular adhesion mechanisms the role of which in normal tissue formation and disease is only beginning to be understood.

1 Edelman GM, Thiery JP, Cunningham BA, eds. *Morphoregulatory molecules.* New York: John Wiley, 1990.
2 Hynes RO. The complexity of platelet adhesion to extracellular matrices. *Thrombosis and Haemostasis* 1991;**66**:40–3.
3 Takeichi M. Cadherins: a molecular family important in selective cell–cell adhesion. *Ann Rev Biochem* 1990;**59**:237–52.
4 Springer TA. Adhesion receptors of the immune system. *Nature* 1990;**346**:425–34.
5 Legan PK, Collins JE, Garrod DR. The molecular biology of desmosomes and hemidesmosomes: "What's in a name?" *BioEssays* 1992;**14**:385–93.
6 Humphries MJ, Olden K, Yamada KM. A synthetic peptide from fibronectin inhibits experimental metastasis of murine melanoma cells. *Science* 1986;**223**:467–9.
7 Gould RJ, Polokoff MA, Friedman PA, Huang TF, Holt JC, Cook JJ, *et al.* Disintegrins: a family of integrin inhibiting proteins from viper venom. *Proc Soc Exp Biol Med* 1990;**195**:168–71.
8 Frixen UH, Behrens J, Sachs M, Eberle G, Voss B, Wards A, *et al.* E-cadherin-mediated cell–cell adhesion prevents invasiveness of human carcinoma cells. *J Cell Biol* 1991;**113**:173–85.
9 Fearon ER, Vogelstein B. A genetic model for colorectal tumorigenesis. *Cell* 1990;**61**:759–67.
10 Berendt AR, Simmons DC, Tansey J, Newbold Cl, Marsh K. Intercellular adhesion molecule-1 in an endothelial cell adhesion receptor for Plasmodium falciparum. *Nature* 1989;**341**:57–9.

Effects of radiations on cells

N E Gillies

Evidence of the biological effects of ionising radiations on cells has been with us throughout this century and for much longer for ultraviolet light. Yet, for obvious reasons, there has never been more interest in the damaging effects of radiations on living cells than there is today. Study of the actions of radiations is not only helping to bring about a better understanding of cell biology but is of considerable practical importance for developments in radiotherapy and in protection against accidental exposure. Radiations may be divided into two main types: ionising radiations, which include *x* rays, cause ionisation of atoms by ejection of electrons; hence their name. This chapter is concerned mainly with these radiations. Non-ionising radiations, of which most wavelengths of ultraviolet light from the sun are examples, are insufficiently energetic to eject electrons but can raise electrons to a higher than normal energy level in atoms.

Relation between dose and response

The most obvious biological effects of radiations on cells are killing, induction of mutations, and conversion to a precancerous state. The causation of cancer by an agent that is well known as a means of treating cancer may at first sight seem paradoxical, but it aptly illustrates the dependence of effect on the dose of radiation applied. Probably only a relatively small dose is required to induce a cell to become cancerous, but for the cell to initiate a tumour it must be capable of continual division over many generations of cells. The high doses used in radiotherapy are designed to prevent that division by killing malignant cells. To the radiation biologist killing usually means that the cell suffers "loss of reproductive capacity" and is incapable of supporting more than a few divisions. The "dead" cell, however, may still be able to sustain virus replication, for example, or synthesise particular proteins; it may therefore retain some biological functions. Of course, if the dose of radiation is high enough all activity is extinguished, and the cell is stopped virtually in its tracks. In other words, an issue of major concern is the relation between the the dose of radiation received and the response of the cell, or more usually a population of cells, to that exposure. This aspect is not always fully appreciated by the lay public, who often think only in terms of an all or none effect.

Much effort has been devoted to measuring the relations of dose to response for the biological effects mentioned. The induction of so called point mutations—that is, changes in the molecular structure of the DNA genetic material—appears to occur at a rate that is directly proportional to the dose. This means that even the smallest dose of radiation has a finite probability of causing such a mutation. In germ cells, however, the expression of that mutation in offspring depends on a range of factors in addition to the molecular lesion initiated by the original damaging event. The rate of induction of carcinogenic change in cells exposed to low doses of radiation is difficult to measure, but methods developed to examine neoplastic transformation in rodent cells in vitro suggest that very small doses may be sufficient to cause malignant change. Furthermore, it is beginning to emerge that there may not be a dose of radiation below which carcinogenic change cannot occur in cells. From the tragic nuclear accident at Chernobyl in 1986 may emerge valuable information to help answer some of the questions concerning the sensitivity of human beings to induction of cancer by radiations. Small doses appear to be relatively less effective than large ones in killing cells. Much debate has been centred on why this should be. It may be that accumulation of damage is required before substantial killing is induced. Alternatively, as now seems more likely, it may be a manifestation of the capacity of cells to repair, within limits, some of the radiation lesions that they suffer. Whatever the explanation, it means that non-lethal damage caused by small doses that may lead to mutation and induction of cancer, for example, may be readily propagated in sublethally damaged cells. Even for killing, the dose of ionising radiation in terms of energy required is remarkably small. As Dr L H Gray, an eminent British radiation biologist, pointed out, the *x* ray energy needed to kill a mammalian cell is roughly equivalent to the heat energy in a cup of tea. The hot brew is not harmful because its energy is not deposited in the same way as ionising radiation, which is deposited in discrete quanta, or packets, of energy capable of disrupting chemical bonds in molecules. Tests in vitro indicate that there is some variation in the radiosensitivity of cells and perhaps especially of tumour cells. In the future it may be possible to exploit this to predict the best dose regimen for the radiotherapy of a particular type of tumour.

How do ionising radiations cause their effects?

After the deposition of energy, how do ionising radiations cause their biological effects in cells? The answer to that question is not clear. Characteristically, the ionisation events are not only discrete but are distributed randomly so that reactions occur in molecules roughly in proportion to their concentration in the cell. The art in answering the question lies in being able to recognise relevant damage in essential molecules, defined as sensitive or critical radiation targets, in a background of "noise" comprising less important or even unimportant biochemical lesions. Professor J A V Butler, a distinguished physical chemist, once suggested that the problem faced by the radiation biologist, intent on unravelling the nature of critical damage in cells, was analogous to that of deciphering why telephones did not work after lobbing bricks through the windows of a telephone exchange. He might have added that the exercise is made even more difficult if knowledge of the way in which the exchange works normally is meagre in the first place. We are facing a tall order, and we are still a long way from a detailed understanding of how ionising radiations kill cells.

Much pertinent information has been unearthed, however, especially in recent years. Firstly, there is no doubt that the critical target(s) are located in the nucleus and not in the cytoplasm of the cell. Again, the most important target in the nucleus is DNA. This must be accepted not only from the unique and central role of DNA or chromatin in the cell but also from the demonstration, before our eyes under the microscope, of chromosomal alterations. Of course, there are molecules other than DNA present in chromosomes, mainly basic proteins, so we must not forget that potentially there are additional target molecules which, if damaged, may contribute to the final biologial response. By means of low energy electron beams of limited but well defined penetration, an area of particularly high sensitivity has been identified immediately within the nuclear membrane. As there is clear evidence of attachment of chromatin to regions of the nuclear membrane, it may be that sites at which DNA and membrane are complexed are peculiarly vulnerable.

Free radicals

Simple chemical experiments in vitro suggest that target molecules in cells are damaged by two routes. One is by direct absorption of radiation energy in the molecule, so that after ejection of an electron the target, which can be represented by the symbol RH, is converted into a free radical. The radical is a chemical species containing an unpaired electron that renders it highly reactive. The ionised target can be shown as R·, the dot delineating the unpaired electron. The second, but indirect, route to the same R· is by reaction of the target with free radicals formed by the action of radiation on molecules that cannot be defined as important targets themselves. Because of its preponderance in cells, water is the main source of these radicals, and one that appears to be particularly potent is the hydroxyl radical (OH·). The view of most workers seems to be that after exposure to *x* rays about three quarters of the initial damage to target(s) arises from indirect interactions with water radicals. The differences between direct and indirect action, however, are becoming less important with the realisation that water molecules are bound to DNA in such a manner that they are virtually part of the structure of the nucleic acid.

The subsequent chemical and biochemical history of the R· species determines the biological fate of the irradiated cell. Within a few milliseconds of its formation R· will react with other available chemical species, which will compete either to fix permanently the damage that has been started or to repair the lesion and restore the structure to its original configuration. For example, if oxygen is present it is capable of rapid reaction with R· by the unpaired electrons that it possesses normally, and the result is the formation of a peroxidised derivative of the target. The damage is now unmodifiably fixed. This mechanism suggests one way in which the well known radiosensitising effect of oxygen can be explained. The other side of the picture is that if hydrogen atoms (H·) are available they can also react with R· and restore it to its original structure.

Chemical compounds that are good donors of hydrogen atoms may be given to cells so that, to some limited extent, they can protect them against the effects of radiation by helping to restore R· to RH. Within the cell itself the tripeptide glutathione, a normal constituent, can also fulfil the role of hydrogen donor. The extent of protection, however, depends on the nature of the chemical environment near the target. For instance, if even small amounts of oxygen are present the competition for R· is strongly in favour of damage fixation and sensitisation because of the high affinity of oxygen for unpaired electron sites. Attempts to make tumour cells more radiosensitive or conversely to render normal tissue cells less sensitive, and so enhance the effectiveness of radiotherapy, often revolve round devising ways of modifying the initial chemical reactions of ionised targets.

Nature of damage caused by radiations

Having identified DNA as a major target, what is the nature of the damage caused in it? Also, of the chemical lesions that are known to occur can any be implicated in the biological damage that is expressed? With the development of new and more sensitive techniques some lesions can now be detected after exposure to doses of radiation that are almost as small as those required to kill cells. These include a range of changes induced in the bases of the nucleic acid, breakage in the continuity of the strands in the double helix, and abnormal cross links formed in the DNA or between it and cellular proteins. It is proving difficult to link these changes conclusively with events leading to killing, but an important lesion is the so called double strand break, in which the structure of the DNA is interrupted at about the same position in both strands. Currently, there is growing appreciation that not only is the type of lesion important but also the number and local distribution of lesions. Albeit the advances in molecular biology are rapidly making the molecular basis of mutation induction, and possibly also of radiation induced carcinogenesis, better understood. It is now possible to insert pieces of "custom built" DNA, containing known molecular changes caused by irradiation, into the genome of cells and then examine the host for the expression of mutations. In this way the link between chemical and biological events may be elucidated.

Repair enzymes

Even if the damage becomes fixed in DNA, all is not lost as far as the cell is concerned. Enzymes capable of eliminating lesions and restoring the integrity of the DNA can be brought into action by the cell. The discovery of repair enzymes less than 30 years ago was a jewel in the crown of radiobiological research and is still perhaps the specialty's greatest contribution to progress in general cell biology. These enzymes, with a range of activities, can locate and eliminate lesions, trim and reconstitute the structure of the DNA, and enable cells to resist radiation and other types of cell injury. Of course, if excessive exposure occurs then these defences will be overcome, but within the limits of their effectiveness repair enzymes fulfil an essential role in cells. This point is made particularly well by the fact that certain genetically determined diseases seem to be associated with patients whose cells are in some way defective in repair enzyme functions. However, as we are only too well aware from our experiences of everyday life, repair work sometimes can be incorrectly carried out: and cells are not immune from such shortcomings. Inaccurate repair of DNA is a recognised phenomenon that can lead to the establishment of mutations in cells. We are probably still only at the beginning of appreciating the complexities of enzymatically controlled repair in cells. Nevertheless, rapid progress is being made on this topic and already several repair enzymes have been cloned, their structures elucidated, and their properties related to their functions.

Further reading

Hall EJ. *Radiobiology for the radiologist.* 3rd ed. Philadelphia: Lippincott, 1988.

Tubiana M, Dutreux J, Wambersie A. *Introduction to radiobiology.* London: Taylor and Francis, 1990.

Index